The FLOWER SHOW

E. D. Wearn

A Guide to Exhibiting Flowers,
Plants, Fruit, Vegetables and
Handicrafts at Local and National Level

CROOM HELM
London · Sydney · Dover, New Hampshire

© 1985 E.D. Wearn
Croom Helm Ltd, Provident House, Burrell Row,
Beckenham, Kent BR3 1AT
Croom Helm Australia, Suite 4, 6th Floor, 64-76 Kippax Street,
Surry Hills, NSW 2010, Australia

British Library Cataloguing in Publication Data
Wearn, E.D.
The flower show.
1. Vegetable gardening   2. Horticultural
exhibitions   3. Flower shows
I. Title
635   SB322

ISBN 0-7099-3620-6

Croom Helm, 51 Washington Street,
Dover, New Hampshire 03820, USA

Library of Congress Cataloging in Publication Data
Wearn, E.D.
The flower show.
Bibliography: p. 140.
Includes index.
1. Flower shows.   2. Horticultural exhibitions.
I. Title.
SB441.W43   1985   745.92′074   85-7770
ISBN 0-7099-3620-6

Typeset by Columns of Reading
Printed and bound in Great Britain by
Biddles Ltd, Guildford and King's Lynn

CONTENTS

CONTENTS

Colour Plates

Figures

Acknowledgements

My warmest thanks to Elisabeth Sheridan-Mills and Leslie Wearn for their help with some of the line drawings and to all the Show Organisers and Exhibitors to whom I spoke, for their ready co-operation in helping me in many ways.

A special word, too, for Charles Jenkins who provided me with a considerable amount of 'garden produce' for photographic use and for Ted Robinson who provided me with helpful literature.

I am most grateful to all of them.

Foreword

In the course of judging at flower shows over the past few years, I have been made aware of a call for a book comprehensively setting out the main elements of concern to the exhibitor and of interest to the spectator. It would not be possible to cover every aspect in detail in a single volume of modest size, but the guidelines in this book will, I hope, satisfy the initial needs of would-be exhibitors and of others who like to visit flower shows; and I would like to think that this book will generally encourage a deeper interest.

I feel very strongly that flower shows are socially desirable events which deserve the widest support. Although they may not appear to offer the same degree of excitement apparent in some other pursuits, they really provide for competition of skills of the highest order with a blend of sportsmanship and a dash of gamesmanship that matches up to any other entertainment. The lure of the flower show has not entirely escaped the attention of the television authorities, and I can recall one or two excellent programmes – such as the National Vegetable Society's inter-branch competition – where the high standards reached and the skill and dedication involved were very well portrayed on the TV screen. Who knows, flower shows might get a wider showing at some time in the future? The more advertisement in aid of flower shows, the more everyone concerned will surely be pleased.

As with all pursuits, there are obstacles to be overcome, not least of which is the task of conveying exhibits to shows and staging them in buildings that are not purpose-built.

There is no gainsaying the fact that the competitiveness of flower shows generates its own brand of excitement, and I hope that this book will introduce many to its delightfulness, be of assistance to would-be exhibitors and of interest to those already involved.

1. Introduction

The 'Flower Show' is a general term for shows of horticultural produce whether it be flower, fruit or vegetable. And indeed, in many shows today there are 'Domestic' classes (that is, cakes, bread, bottled fruit, etc.) and 'Handicrafts' covering a wide range of artistic and technical skills such as painting, photography, needlework and so on. In some shows Home-made Wine has its own well-supported section; and, of course, 'Flower Arrangement' or 'Floral Art' is an essential part of many shows, attracting usually a largish number of entries. In fact, I find at a number of shows that the Flower Arrangements and the Domestic items comprise a predominant proportion of the total number of exhibits.

Flower shows have a long history and in earlier centuries they were an event of much importance to village community life. The village shows, as they were known, provided a highlight to the year and the forum at which folk from neighbouring villages could renew acquaintance annually. Today, it is pleasing that the term 'Village Show' is still in use – even in some London suburbs – but the character of the many thousands of flower shows held annually (with some societies holding two or three shows a year) varies very considerably as does the support for them.

No doubt the change in life-style has had a big impact on the more simple pleasures derived from supporting or partaking in communal activities; and horticultural pursuits have to fit in with other commitments which nowadays at weekends seem to be far-ranging. To attract family participation in flower shows, committees have to think of ideas that will appeal to all age groups, and the show schedules must be comprehensive in this respect. One idea which I have seen to be successfully employed is the distribution of seed of a fairly easily-grown but interesting plant accompanied by a moral obligation on the society members receiving the seed to enter the forthcoming show, the schedule for which will contain a special class for the purpose. The not-so-humble pumpkin is a good choice, and more often than not children welcome the opportunity to compete with their parents in producing mammoth specimens. As well as being fun, it provides a useful challenge for the family to take up: and, of course, it can be the 'starter' for a deeper interest in supporting the local society and entering into the fascination of exhibiting.

Whilst the bulk of flower shows comprises local or village shows held in village, school and other public halls or in marquees on the local common and suchlike venues, there are – arguably at the other end of the scale – national shows such as those organised by the Royal Horticultural Society at their Halls in Vincent Square, London, and those run by National Societies who often stage their shows there. The support for these shows varies, of course, but even if not overwhelming, a fine display of high-quality exhibits is usually to be seen. There are also Borough and County Shows which provide 'all the fun of the fair' with a combination of horticultural exhibits, sideshows, dog shows, gymkhanas and many other features that cater for everyone. And in writing of flower shows, one can hardly omit reference to the 'Chelsea Show' of world renown and held annually in London. The underlying theme at 'Chelsea' is to produce a magnificent spectacle (see Plate 2) and the nurserymen who exhibit vie for the coveted Gold Medal which is awarded to exhibitors for outstanding displays. But there is no section for amateur competition as there is at some other large-scale shows, including the famous Southport and Shrewsbury Shows which, like 'Chelsea', are annual events.

I think it would be difficult to go to any flower show and find nothing of merit, and at local shows there are usually some exhibits of the highest quality because aspiring national-level exhibitors will win their spurs there and many who have already won fame at the higher level will continue to support their local show. But, of course, one would not expect the overall standard at local level to rival that at national level; yet that is of little matter as far as the enjoyment factor is concerned; and as far as the exhibitor is concerned, being the local champion has its own sense of pride. And that somehow brings me to the aspect of prize money. Basically, most exhibitors are classed as amateurs and it is rare that they can win much more than could be regarded as very modest sums, although some societies have special fund-raising activities and, coupled with prizes in kind donated by local stores or firms, this can add a touch of material gain to winning. There is, however, some movement in financial sponsorship by commercial enterprises – although, of course, not on the scale applicable to golf or tennis! Some of the most attractively financed shows are pot leek and onion championships, and intense competition is guaranteed. The exhibits of staggering majesty have a special fascination for the public and there seems no reason to cavil at such events. On the other hand, to win the first-prize card – particularly against stiff competition – is, I hope, the main inspiration to most exhibitors, and it can be saddening, I feel, sometimes to hear the remark 'it is not worth [financially] entering'. I suspect that it has never been a very profitable business to compete in flower shows if the criterion were solely money-based – and that the real gain was in pride and enjoyment.

The field then is very wide; not only are there opportunities in local shows to display one's prowess in a fairly wide range of skills, but there are other goals at which to aim such as making one's mark at national level or, for that matter, becoming a very knowledgeable visitor to shows. And individual tastes will vary, some deriving joy from the exquisite beauty of tiny alpine plants and some seeing a classical display of near-perfect exhibits of fruit as out-rivalling everything else. Whatever the choice or the depth of interest, enjoyment should be the aim both of the exhibitor and the spectator.

Information about flower shows is available in many ways: members of local societies will know all about their own shows and they will spread the news around. Usually, too, the shows are advertised in the local press. Some gardening weeklies publish details of all shows during the year and, of course, the national press provide information about the national shows and other major events. All in all, it should not be too difficult to get the information you require, and there is always someone, if not an immediate neighbour, who will cheerfully let you have the necessary particulars.

A Note for Non-UK Readers

Broadly speaking, access to details of shows will similarly be available through membership of clubs and societies, and from information in the media. See also Appendix V.

The months mentioned in the text are those applicable to the United Kingdom; the following is a rough guide to appropriate weather conditions. The average UK daily temperatures for February-April rise from a minimum of 2°C (35°F)–5°C (40°F) to a maximum of 7°C (45°F)–13°C (55°F). In May-August, the averages would be 7°C (45°F) –13°C (55°F) to 16°C (60°F)–21°C (70°F). In September, the range would be about 10°C (50°F)–18°C (65°F). Average monthly rainfall is approximately 5 cm (2 in), but 7.5 cm (3 in) is not unusual during certain times of the year.

However, dry spells of a week or more can sometimes result in parched soil during the summer months of June, July and August.

Most of the fertilizers mentioned in the text are available outside the United Kingdom. The two compound fertilizers used in solid form – fish, blood and bone, and Growmore – may not be available overseas under those names, but similar compounds are available. Both types have an even balance of nitrogen, phosphorus and potassium; the former, however, having an organic base, is biased slightly towards potassium. Of the two that I consider very useful for liquid application, Maxicrop is of seaweed extract, is supplied as a liquid and is available in the USA and elsewhere; and Phostrogen, which is easily diluted but can also be used in its solid form, is of chemical make-up and is widely available.

2. Getting the Feel of Shows

The layout of shows varies tremendously depending upon the facilities available. Usually trestle-type tables are set out in rows along the hall or marquee, often in a rectangular arrangement. More sophisticated arrangements do not necessarily make a better show apropos quality and number of exhibits, but a covering for the tables of hessian or baize helps towards an attractive display and a backboard – as provided at major shows – enables certain kinds of vegetables when exhibited in collections to be staged in an upright position thus giving an enhanced appearance to the show bench. Also, niches made from corrugated cardboard or similar material are often provided for floral art exhibits and the exhibitor will be able to add a silk or rayon drape, if that is permitted in the show schedule. The other main require-ment is to mark the class divisions by lengths of tape or ribbon – the term 'class' denoting each item in the schedule (see Appendix III) which is the document that lists and describes all classes and sets out the rules for the show.

The minimum amount of guidance on the show bench will be the number of the class, so that it can be related to the description in the schedule, under which the particular exhibits are entered for the show. More elaborate arrange-ments include small notices with both the number and a brief note of the class and sometimes these are surmounted by larger notices denoting class divisions or sections. Whatever the arrangement, a copy of the show schedule is not only a necessity for the exhibitor but also a useful aid for the spectator.

There are many who visit shows and are surprised to see that the exhibits appear to be no more than on a par with what is growing in their own gardens, or what they may have baked that very week, or with the photographs they have just had processed. The trouble is that they have left it all at home and, therefore, it has no chance of proving its merit; and some of the hesitation in entering it for the show may well have resulted from doubt concerning the rules and requirements. So before perambulating round the show, why not obtain a copy of the schedule so that you can compare the requirement therein with the exhibits and become more aware of what is required? The odds are that you will find it more simple than you thought and that little breakthrough will be sufficient encouragement to persuade you to have a shot next time. There may be some difficulty in obtaining a copy of the schedule if you are not a member of the society sponsoring the show, but at least try to borrow one from a committee member or steward; and very often spare copies will be available on the committee table. Becoming familiar with show schedules is an essential ingredient for exhibiting and for the spectator it will add to the enjoyment of the occasion.

Having armed oneself with the show schedule, there is no need to rush in and study each exhibit in strict sequence, and the size of the crowd may make this tiresome. If one has time it is no bad thing to sit somewhere, hopefully with a cup of tea, and peruse the schedule to see what general rules obtain. Then, rather as the judges will have done when commencing their deliberations, a general survey of what the show has to offer will enable you to decide where to start – or, of course, you may have some personal preference for certain sections of the show. May I suppose that you are particularly interested in the fruit section and that you make towards it? The next discipline is to look at the scheduled requirements first and then the exhibit rather than the other way around because that habit needs to be acquired when starting to compete.

Assuming that you are new to the idea of examining in depth exhibits at shows and the first class in the fruit section is for a simple exhibit of three eating apples, the first thing that may seem odd is that the fruits will be sitting on their tails, so to speak, but this is the normal method by which they are staged (see Plate 17). The object now should be to pit your wits against the judges', and a conscious effort should be made to avoid being influenced by the awards on the exhibitors' cards which will very likely be coloured stickers carrying the words 'First', 'Second' and so on. Fourth position, incidentally, usually gains a 'Commended' rating. If there are six exhibits in all, an effort should be made to eliminate in your mind two or three of the exhibits that you consider to be inferior to the others. There may not be any obvious blemishes to help you in this task, but evenness of size of each apple in the three representing each exhibit, or 'dish' as it is called, good colour matching and uniform shape should be sought, and on this basis it is not very likely that all six exhibits will have the same appeal. Having chosen what appear to be the best three or four dishes by eliminating the others, try to put them in order of merit as the judges will have done. Throughout, of course, they will have carried out a more rigorous examination than you are able to make because they will have been entitled to handle the fuits and look very closely at them, but matter not, it is a valuable exercise to test your powers of observation albeit at a slightly longer range than if you were actually judging. Of course, your selections may well not tally with those of the judges; on the other hand, you may be pleasantly surprised by the degree of 'unanimity' between your ideas about the merit of the exhibits and the views of the judges. If all this encourages you to follow up on these matters, a great deal of information on the judging of all classes is contained in the Royal Horticultural Society's *Show Handbook* (available from them for a modest sum). A copy will be a necessary tool for you later on when exhibiting or judging.

The main thing is to move round the show enjoying it, and rather than making an intense study of each class in turn it could be more congenial to pick and choose here and there covering perhaps all the sections or divisions by choosing one or two classes in each one for particular study. But before leaving the Fruit section, a look at any soft fruit such as strawberries or raspberries will underline the fact that freshness of produce plays a key part in its selection for prize awards. It will be seen that some dishes of fruit contain specimens that fairly glisten wheareas others may look dull and tired. Then again, among the healthier-looking dishes of, say, the strawberries there will be a greater degree of uniformity of shape and size in one dish compared with another. In simple terms, the dish containing the winning combination of perfect freshness and good uniformity will be placed ahead of another dish of equally fresh fruits which are ill-matched in size and/or shape. Or, of course, a dish containing some very over-ripe fruits would lose to the latter despite the gap in uniformity. And so it is that throughout the show, freshness, colour, size, shape and uniformity will play their part in the award of prizes.

Moving round the show and coming to the flowers, let us suppose that a halt is made at a class, so popular in summer shows, for a single delphinium spike. There may be several entries and among them a loftier spike than the others that has nevertheless not won first prize. It is difficult at this point to follow the advice that you should, having studied the schedule, do your shadow judging uninfluenced by the markings on the cards, but it is no bad thing to look for reasons why the towering bloom has been considered to be of less merit than two others of smaller overall stature. Perhaps it has a nasty bend at the top, or the florets are unevenly spaced, or some of them are slightly damaged? The two spikes winning first and second places, however, might be upright and nicely tapering with no large gaps between any of the florets. So you may feel satisfied that there is substantial reason for them to oust the heavyweight, but it may now be a little harder to determine what has caused one of the two to be placed above the other in order of merit. It will then be a matter of looking for any signs of lack of freshness in some of the florets or even a few snags in some petals. It may in fact be easier to see that the two smaller spikes should beat the larger one than to decide which of them should be placed first. Thus the task of judging finer points is difficult but as far as learning about showing is concerned, it is important to note that size has to be accompanied by quality if it is to be a virtue.

An interesting section to test one's powers of observation is that for Pot Plants, and usually there will be separate classes for flowering plants and foliage plants. The intention of the schedule-maker here is that the former should have their blooms predominating, although they should be accompanied by healthy foliage. Foliage plants, on the contrary, should be shown for their fine display of stems and leaves, and flowers if any would be expected to be insignificant. And, should the schedule say 'foliage only', flowers should be absent. In all cases, a healthy plant is a prerequisite and it has to carry a good showing of blooms if entered in the flowering class or an especially attractive head of foliage if in that class. Unfortunately, the first thing that may be spotted on the bench is a card with the initials 'NAS' (Not as Schedule), and it is usual in these cases for the judges to put a brief explanation of the reason for this. Several possibilities obtain, one of the most frequent of which is that the pot exceeds the stated measurements. Usually the schedule states a maximum size, but sometimes – and this applies at shows run by national societies where a standard format of presentation is desirable – it is more specific. Whatever the requirement, the exhibitor must comply, and it just needs that little touch of self-discipline to avoid the disappointment of being adjudged 'Not as Schedule'.

While on the subject of the pot itself, it is of importance that it should be clean, and in studying the exhibits on the bench it might be found that the only obvious difference between two very nice plants is that one of them is in a grubby pot. If, in fact, the plants are, as far as can be judged, of equal merit, the dirty pot may well indeed have been the deciding factor.

I hope that the idea of going round the show comparing schedule with exhibits and studying some of the exhibits in depth will be of interest and add to the enjoyment, as I am sure it already does for many; and I think there is little doubt that it will greatly assist those who are keen to make a successful start to exhibiting. Remember though that enjoyment is the keyword and whatever the approach to becoming more involved in shows, the effort must not be felt to be tiresome.

The extent to which one eventually becomes involved in flower shows will obviously vary from individual to individual as will also the rate at which the ladder to success is climbed. Some exhibitors have commenced at national level and many others have preferred to make haste slowly by becoming accomplished at local level over a period of some years and then deciding that they ought to have a go at winning coveted awards at national shows. There is much to be said for serving an apprenticeship for two or three years before tackling national shows, if only because one would want to avoid making elementary errors there. A lot depends upon one's personal rate of learning and the time that one is likely to have available. In general terms, however, a good plan would be to be a keen 'observer' at shows for a year and in the following two years learn more about exhibiting by entering local shows; and then, if not content to aim at being the best local 'all-rounder', decide upon one's further horizon. Hopefully, too, there will be many who will also find it worthwhile to assist local committees in preparing shows – which is a very good way in which to find out more on the whole subject.

3. Producing the Exhibits

There are some keen gardeners who think that it is a waste of effort and space to grow flowers; for them, vegetables and fruit are sensible choices with a practical value. Others regard vegetables as mundane objects, and many others like all-round gardening, involving growing flowers for colour in the house and garden, fruit and vegetables for culinary use and perhaps maintaining a lawn, with additional features such as rockeries and ponds. In making an initial entry into exhibiting in the horticultural sections of a flower show, however, it is advisable not to over-stretch, and some rationalisation of one's gardening effort will become necessary. Organisers of shows naturally want as many exhibits as possible, and it is nice to oblige in this respect: but they also want exhibitors who stay the course and can be relied upon to produce some good exhibits every year. So it will not be helpful to anyone if the exhibitor's entry into the scene is so hectic that he or she becomes exhausted after the first attempt and consequently goes into decline. A determined beginner and not an over-enthusiastic 'trier' is the likely winner in the long run, so a modest and purposeful start should be the aim.

In looking for areas in which one might expect to do reasonably well at shows, it has to be assessed what potential exists, and this is governed, of course, by physical limitations such as size of garden and its position and type of soil, or perhaps the size of balcony on which everything may have to be grown in pots. And additional facilities in the way of greenhouses or, in the case of those interested in handicrafts, workrooms or even studios, will need to be taken into consideration. Naturally, too, one's own abilities are an important factor, but all in all, it ought to be possible to decide on one or two areas in which to make a start: and from there a broader field can be tackled within a year or two during which it should have been possible to have had some success – that much-needed boost to one's confidence!

Indoor Gardening

Starting with a limited situation such as a flat-dweller with no outside gardening facility at all, the only real possibility in the horticultural section is to grow suitable pot plants indoors, and for flowering plants I suggest African violets. For a good foliage plant that will not tax one's skill too much, the plant known as mother-in law's tongue (Sansevieria) would be suitable despite the fact that because of its nick-name it is treated as rather a joke. Or another with a music-hall touch – the aspidistra – should by no means be ignored: and, in fact, a goodly specimen might need to be guarded because of the value that is currently put upon some of these 'Victorian' plants. The fact is that both of these foliage plants will put up with variable conditions in the home much better than many others, and the African violets can also be accommodating, although, of course, they should be treated kindly if they are to be expected to win prizes for you. I have put some notes on cultivation and preparation for the show in Chapter 13 'Pot Plants'.

For those with no outdoor gardening facility wishing to exhibit more than just one or two pot plants, there is the possibility of entering the Floral Art or Flower Arrangement section for which flowers do not usually have to be grown by the exhibitor (check on this, however, with the show schedule). This is a time-consuming choice and it might be found that not very much else can be tackled for a start, although it should not add a lot in effort to also exhibit the pot plants. For flat-dwellers who have a balcony or verandah, the field is opened wider and provided that the load-bearing capacity is known, there is every possibility of growing a variety of fruit and vegetables in large pots in balcony areas. Hopefully gale-force winds do not sweep the site, and the major problem then will be keeping the pots from drying-out in summer. A good mulching of moist peat will be necessary, and the surface of the pot can be protected by polythene or a layer of shingle although the former might be found to cause the soil to heat up too much. Certainly tomatoes are a real possibility on a balcony, and dwarf fruit trees in pots of some 45 cm (18 in) in diameter and depth with a minimum of 37 cm (15 in) containing some nourishing medium – for example, a mixture of good loam and composted organic material – with some broken crocks at the bottom – can do very well there if they are re-potted in the autumn every three years and root-pruned on the lines of Bonsai treatment.

Outdoor Gardening

Coming on to those with a garden, we find, of course, that a wide range of plants can be grown or is already being grown and, as I have already mentioned, it becomes a matter of making a choice of what to exhibit. It is possible to become a keen exhibitor by getting the show schedules each year and making a selection from what is available in the garden at the given times, and summer shows benefit greatly from this. But I think that it adds a touch of spice to select a small range of plants upon which to devote just a little extra loving care so that they are of higher quality than if left to their own devices, so to speak.

There are few plants among those commonly grown that will not show their appreciation of a little extra nourishment, although I would emphasise that over-indulgence is to be avoided and, of course, a lot depends upon the soil fertility that is present. There are, for example, exhibitors of vegetables who get their top-class specimens by growing them in beds of soil that have a very high humus content brought about by regular applications of composted manure; further 'feeding' is shunned, the only concession perhaps being a light application of sulphate of potash to assist ripening. A look at what is flourishing in nearby gardens and comparing progress of like plants will serve as an indicator of how well you are doing, and if there is no cause for disappointment, all that may be necessary to produce absolute 'winners' could be an application of a general-purpose fertilizer such as Growmore or fish, blood and bone – which I think is preferable because it has an organic base – in the spring. Further improvement could be brought about by fortnightly applications of a liquid feed such as Maxicrop (seaweed-based) or Phostrogen from, say, the end of April until early July.

What I am suggesting so far is a modest extension of your gardening involvement by taking one or two fairly straightforward steps to improve the quality of your produce and, therefore, enhance its chances at the show – and indeed its image. The selection of plants for special treatment as part of your plan to become a successful exhibitor will depend upon a number of factors, not least of which is personal preference for particular kinds of fruit, vegetables or flowers – I do feel that one has to like the selected subject in order to bestow the right amount of care on it! In any case, it is wise to restrict oneself to what is likely to be feasible rather than be overwhelmed by sheer enthusiasm; as I have mentioned earlier, whilst show secretaries wish to see halls or tents brimming with flowers, they also want to see you entering year after year and not fading from the scene after one dramatic entry into it. The second main factor concerns the environment in which everything is grown – whether the garden or part of it is heavily shaded or exposed totally in summer to blazing sun. In the latter situation, it would be well worth considering growing some giant zinnias which one seldom sees at shows, or if a woodland area or similar shaded garden predominates, why not grow some fine foxgloves of the 'Excelsior' type? Whatever the choice – and, after all, it might be a burning ambition to show dishes of fruit to out-rival some of those seen at shows that you have attended – it should obviously be such that it is compatible with your gardening environment or ability – or else frustration will result!

Of course, there is the alternative of altering everything in the garden to match one's aim, but such decisions require a great deal of thought. There is always some scope, after your 'familiarisation' visits, for doing something about entering the next show without massive changes in the normal pattern of life, and indeed whilst success might not come too easily, it does not have to be gained at the expense of all else. I hope that some of the suggestions made in a later section on selection of produce and other items for exhibition will help in this respect.

In many ways a piece of land away from the home is an extremely useful asset, although it might lead to transport difficulties when the time comes to harvest produce for the show. An allotment site, for example, is usually an ideal place on which to grow soft fruit and vegetables. Watering can be a problem but this is somewhat compensated by the fact that large loads of manure can be deposited on the plot bringing about a good humus content in the course of time; something that is often needed in the garden but difficult to achieve for obvious reasons! Top-level chrysanthemums and roses are grown for exhibition on allotments, but one usually thinks in terms of growing fruit and vegetables when taking on an allotment – which indeed is usually its real purpose – and it might be found quite sufficient to concentrate on those if exhibiting high-quality produce is in mind. Oddly enough, many allotment holders grow exhibition-quality vegetables yet never wish to enter them in a local show. I have often thought that, if persuaded to enter on any one occasion, many would be bitten by the bug and would find a new enjoyment. To some extent I have proved this in assisting colleagues to select some vegetables from their plots. On finding that, to their surprise, they have won an award, they have accordingly derived much pleasure from it: this has thereby whetted a few appetites for 'the game'.

In parallel with the idea that you tailor your exhibiting aims to suit your potential in terms of time, facilities and abilities, goes the idea that I like to propound regarding perfecting your produce for the show: in short, it is possible to reduce the effort involved by a little pre-planning. For example, many of us have been involved in what I term spasmodic 'spraying' rather than systematically setting about it. Yet it has probably been just as wearisome. Would it not be wise, for example, to give the rose beds a good clean-up in the winter with a suitably diluted (according to the instructions of course) wash of tar oil or Jeyes fluid in order to provide a clean start in the folowing season, and would this perhaps avoid an early spray of an insecticide?

In propounding this theory, I am also conscious of the fact that as far as insecticides are concerned, it is no bad thing to restrict their use as much as possible, and to employ non-persistent kinds where practicable. I realise that it is useful when going on holiday to give plants a dose of a systemic spray in the expectation that during that two or three weeks blackfly or greenfly will not take charge; and I doubt whether a perfect programme of biological control is within the scope of the majority of us. But where there is a choice, it would be wise generally to use an insecticide that ensures a quick kill of the foe with minimum residual contamination; and in all cases, the maker's instructions must be read thoroughly. (Information is available in booklets written by the Ministry of Agriculture, and the Henry Doubleday Research Organisation specialise in biological control.)

In producing exhibits for a show, therefore, no matter which section of the schedule is of particular appeal, a certain amount of effort is necessary, but by no means need it be other than enjoyable. A little planning to take the maximum advantage of existing potential is better than becoming over-involved too quickly; and enjoyment is the essence that leads to a long association with the subject.

4. Harvesting the Produce

There is no gainsaying the fact that having grown good produce it needs careful handling if it is to look its best on the show bench. In commercial harvesting, some damage is inevitable although the growers would prefer to avoid it if possible: to the exhibitor, however, it spells disaster. Fortunately it only requires some thoughtful care to avoid spoiling your exhibits and, after all, this is a fairly automatic step taken by anyone entering a sponge sandwich or some fancy cakes. They would be safely packed in a tin with the minimum of handling and would be expected to be in the same excellent shape when displayed on the bench as when taken from the oven; and that should be the aim for the exhibitor of horticultural produce although different considerations arise.

Cut Flowers

In general, cut flowers deteriorate all the time they are out of water although, as with everything, there are some exceptions to this 'rule' – with which we will not concern ourselves here. An initial deep drink pays dividends as the flower arrangers will avow, and it follows that the best arrangement is to have containers of water on site with you when cutting for the show. A cut at a slant of 45° (Figure 4.1) will provide the maximum surface at the stem end to come into contact with the water and will also ensure that there is no adhesion to the bottom of the container. Make the cut with a sharp knife or a sharp pair of secateurs or scissors, depending upon the toughness of the stem. Plastic buckets will not normally tip over when filled with water but, of course, flowers such as delphiniums will need to be placed in a deeper container than a small bucket otherwise they will topple – and it would serve badly to prop the blooms against the fence or wall. Small beer casks would be useful if they can be obtained, or you might obtain large tins from builders' yards. These tins will have contained various powdered substances which can easily be washed out. Although it is not invariably necessary, some exhibitors always cut stems under water, having first secured them from the plant, because air enters the stem once it is cut and in some cases this will cause an airlock so preventing any water reaching the bloom. The relevant flowers are delphiniums and the like with hollow stems which can obviously quickly fill with air, and the best idea is to fill the stem – held carefully upside down – with water from a small can such as the type used for indoor plants, and then plug the stem end with cotton wool, thus ensuring a stem filled with water. This is then placed in the container of cold water as with other flowers.

Figure 4.1
Flower stems should be cut on a 45°slant, and woody stems should be slit

Other special measures are to dip the ends of woody stems from shrubs or trees, such as lilac, briefly into boiling water in order to reduce callusing and with poppies and other plants of similar nature which exude a milky fluid (latex) when cut, to cauterise the stem end by quickly passing it through the flame of a match. Hard stems ought to be slit for 2.5 cm (1 in) or so at the bottom and that amount of bark can usefully be scraped from the ends of really woody stems. In all cases and after any special treatment, the stems must go into cold water which should reach to just below the bloom or lowest floret, but the blooms should not get wet. Of course, weather conditions are not always perfect and there are times when it is safer to cut the flowers and take them indoors where it is dry, without trying to put them into containers in the garden where they might get tossed around by gusty winds. Cutting the stems again under water would then be appropriate.

Having cut the flowers and placed them in water, they should be left as long as possible until being packed for the show or, in some cases, it might be possible to take the containers as they are and so avoid extra handling of the blooms – but more on that in the next chapter dealing with 'getting to the show'. The important thing is to keep the flowers in a cool place and to try to ensure that where multi-headed spikes are concerned, they are not placed in a position where they reach towards a distant window thereby producing a nasty curl to the tips.

One key factor in all this preparation is when to cut the blooms. A lot will depend upon when the show schedule will let you stage them (prepare them for exhibiting) and the time at your own disposal, bearing in mind how long the journey from home to show will be. At County, Borough and National Shows, it is usual to find exhibitors staging into the night and again very early in the morning, but at local shows a lot of the staging takes place in a few hours allotted on the day of the show – say, from 8–11 a.m., although many provide for additional hours on the eve of the show. In general, it is best to avoid leaving everything until the day of the show even if it is 'local'. Flowers such as pansies, for example, are likely to wilt in their shallow containers at the show if they have not been prepared by being given a good long drink beforehand, and a great deal of time is needed to clean vegetables. To do it all in a rush may well result in hasty staging and consequent lack of success. So, although you might look into the garden at dawn and spot one or two 'winners' among the roses – which indeed you would be foolish to ignore – I think that probably the eve of the show will be more generally suitable for cutting the bulk of the flowers, and indeed, there are some kinds which will even keep fresh for a little longer (see Chapter 10).

You will, of course, work out your own plan to suit your individual capabilities. There are many who think little of travelling hundreds of miles to reach a show venue in the very early morning so that they then have perhaps three hours in which to complete their staging, but there are others who prefer to do it all rather more slowly the day before. Too much rushing can lead to disappointment. A cool, calm approach is required – including how to keep the flowers fresh, bearing in mind how and when they need to be transported to the show – if indecent haste in staging them is to be avoided; and this will come with experience.

In selecting flowers for cutting, it is important to set yourself a high standard. If flowers look past their best they must be ignored; nothing will improve them and by the time of staging they will embarrass you. It is a good idea to refer to the copy list of entries made when submitting your entry form to the show secretary; and check with the schedule to make sure of the requirements for the classes in which you are entering. If one of the classes entered is, say, for six blooms or stems, it is to be hoped that you will find rather more in good condition so that a small reserve can be taken to the show as an insurance against damage in transit or during staging. This should be modest in number because otherwise a time-consuming effort will be required at the show to sort out the best. For six blooms or stems, I would suggest taking eight; and for twelve, a reserve of three would normally be adequate. Apart from the prime need for freshness, a criterion of high importance is 'Uniformity' – both of size and of colour, and also of form or shape. Whatever the exhibit, when a quantity greater than one is involved, the judges will be influenced by the balanced effect that is achieved; unless you are desperately short of choice, you will have to make some hard decisions such as overlooking a majestic bloom or spike in favour of a smaller one that, however, matches well with others to provide a uniform exhibit. There are pointing systems set out in the RHS Show Handbook and in National Societies' Handbooks by which exhibits can be assessed, and 'Uniformity' is among the items listed for flowers, and for fruit and vegetables too; for flowers, it caters for about 15 per cent of total points. In practice, however, pointing is the exception rather than the rule and as, following freshness, what impresses next is balance or uniformity, I think that it probably weighs with a judge more heavily than the percentage basis I quote. So it is well worth taking time to select well-matched blooms or stems.

Fruit

Whereas it may seem natural to avoid handling flowers other than by their stems a more conscious effort is needed, I think, to prevent oneself taking hold of fruits in the palm of the hand or clutching them with your fingers. Yet it is of paramount importance to touch fruit as little as possible, and it should be held by the stem or stalk with gentle fingertip support where necessary from underneath (Figure 4.2). With practice, this can become 'second nature', and the exhibits will benefit thereby. It follows that if especial care is to be taken to avoid damage to fruit by unnecessary handling, it will need to be exercised also when placing fruit into containers. Thus trugs or boxes should be lined with soft tissue paper or even cotton wool and it will be useful to have these to hand, so that the fruits can be gently lain into them as soon as they are picked. As all fruits are expected to be shown with stalks intact, with the exception that peaches and nectarines have minuscule ones and therefore escape the general rule, it is necessary to secure soft fruits from the plant or bush by cutting the stalks at their far ends with a sharp pair of long thin-bladed scissors. Soft fruits are described as those having a soft texture and numerous seeds

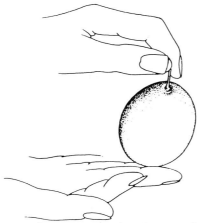

Figure 4.2
Fruit should be held
carefully

and include strawberries, gooseberries, raspberries and currants, and although I know of practised exhibitors who can secure the first-named by a dexterous use of a sharp thumbnail, I would not recommend this method to beginners.

In the case of 'top fruit' such as apples or pears, it is well known that when ripe the stalk should part from the tree if the fruit is palmed gently upwards and given a slight twist and this is found to be quite satisfactory by many exhibitors, but apart from the fact that some pears need to be picked at a stage which allows them to ripen over a short period off the tree, there are such difficulties as finding yourself trying to pick the best-looking fruits – perhaps apples – by standing on a stepladder. In this situation, a good idea is to cut the piece of wood to which the stalk is joined and carry this as an entity onto *terra firma* when the stalk can be carefully separated from the wood. Whatever happens, the stalk must remain attached to the fruit. A little practice with some apples that are not required for showing will help.

In selecting the fruits with the most potential, freshness and uniformity of shape, size and colour will be the main guidelines, and a reserve number should be 'picked': say, an extra apple or pear for each three to be exhibited and an extra quantity of about three soft fruits for each exhibit of ten in number. Weather conditions will play their part in not only timing of the picking operation but in the method employed. Apples, for example, will keep their freshness for at least a few days if stored in a cool place and, therefore, if rough and windy conditions threaten it would be wise to pick them before the weather strikes. Whilst gooseberries will last a few days in a cool room without showing too many signs of tiredness, other soft fruits will stay fresh for only a short time, and really twenty-four hours is the maximum period that they should have to spend off the plant if they are to look their best. Raspberries and strawberries, for example, need to have their glistening appearance preserved for the judges to admire and they will lose favour if they look tired and wan. I hestitate to give any advice about keeping fruits in a refrigerator, and there are some exhibitors who take a simple line that any such artificial aid is simply not worthwhile. But there are others who find that resorting to such methods is often necessary – not only for fruit, of course – and that no real harm is done if refrigeration is practised for only short periods of time, say up to forty-eight hours. One problem with refrigeration is avoiding condensation; a difficult task, I find. Some experimentation is required to determine the most favourable spot in the refrigerator and the most suitable container for the fruit – bearing in mind that it is best to avoid too much handling and that desirably the fruit should be placed in a tissue paper-lined box or something similar and not handled again until it is staged. And if condensation forms on the fruits – as I invariably find to some degree – it must be allowed to evaporate naturally rather than being mopped up with a vigorous dabbing with a cloth, although the more solid fruits of gooseberries and the like will accept a gentle wipe with a very soft cloth.

It is all very much a matter of compromising sometimes between the ideal and the practicable, and it has to be accepted that weather can play a large part in the whole business. If, for example, one has to sally forth onto an allotment plot in wet weather to pick raspberries and currants, it could be very wise to gather the crop by cutting some of the fruited wood complete with its bunches of fruits and take it to a cool dry place at home where the task of securing the individual berries or strigs can be completed in comfort. The mini-branches or canes can be stood upright in vases or similar containers and if time permits they can be left to dry a little before you start to work with the scissors; nearby artificial heat sources should be avoided.

The important message to bear in mind at all times is that good fruit can so easily be spoiled by rough handling, and holding the chosen specimens by their stems or stalks is better than clutching them in the palm of the hand. But, of course, the weight of the fruits must be considered, lest the stalks tear away; so, with large ripe plums, for example, a careful lift underneath with two fingertips would be wise. I see advantage in wearing a pair of thin cotton gloves for this sort of job and for handling peaches and similar delicate fruits. Peaches and nectarines have practically no stalk, and a safe way to harvest them is to cut away the whole woody shoot to which minuscule stalks are joined and then remove the former carefully when sitting comfortably at a table. Grapes offer an obvious method of harvesting in that they have to be exhibited not only on a stalk but also with a piece of shoot so that the two form a T-shaped handle which is used to suspend the bunch on a small stand when staging (see also Chapter 6). The bloom on grapes – the fine dust-like patina that makes them so attractive – must not be finger-marked, and thus the handle of wood that is cut when harvesting is very useful. Melons are also harvested with a piece of shoot being retained, and they are fruits that can improve in quality after being cut from the vine, the difficulty being the matter of judging the optimum time for that operation. The usual method of testing for ripeness is to press the flower end of the fruit gently with the thumb, and if there is a distinct feeling that it is giving way to this light pressure, the fruit is reaching perfection. A few days in a cool place can then help because the fruit will continue to ripen to the point when it exudes the juicy smell that judges seek.

The timing of cutting or picking fruit for a show is, therefore, a moveable feast, and with fruits that we can expect to complete a perfect ripening off the plant, there is the advantage that part of the preparation for the show can be dealt with in leisurely fashion 'earlier in the week'. The melon is a case in point, and some pears also fall within this category. Broadly speaking, early cultivars can be harvested before they have changed colour to indicate a fully ripe stage, and a good plan of campaign would be to aim at picking them over a period of eight to five days before the day of the show, with the expectation that some will colour-up nicely in cool storage. Those that look fully ripe on the tree are best left until a day or so before the show. Most effective, however, is knowing the habit of each cultivar being grown. And as the year advances into its later stages, weather will have to be taken into account: then again, there are some 'late shows' for which fruits will probably have to be stored a while, the emphasis then being on keeping them in a cool place to preserve their qualities. Given suitable weather and date of show, however, late cultivars of pears are often left on the tree until they are about to part naturally; if picked too soon they will quickly shrivel. Matching colour is, of course, desirable, and it can be possible to pick fruits one or two at a time over a period of about a week so that they acquire similar tones or shades in storage. I have known them placed for a brief spell in a biscuit tin in a warm place, but this, I think, is a judgement best left to those with experience, and generally a warm room is to be shunned.

Apples are more tolerant regarding 'picking time', but a customary practice is to leave early cultivars on the tree until ready for eating. They can, however, be picked at least a few days before the show. Some mid-season cultivars, of which 'Ellison's Orange' is one, can benefit from being stored for several days, and in general – with an eye on the weather –

late cultivars would be left on the tree as long as is considered 'safe'. Cooking apples are dealt with similarly. Needless to say, warm storage is totally wrong.

The aims in harvesting fruit are therefore simple: to gather it at just about its moment of perfection as near to show day as possible, and then to preserve it in its pristine condition, which means handling it as little as possible and packing into padded containers with added protection from cotton wool and tissue paper. That 'best moment' for picking, however, has to be judged in the light of the habit of the kind of fruit and cultivar involved, the weather and personal experience. With many fruits and for most shows other than perhaps late ones, picking a day or two, and not longer, before a show would be of advantage; and soft fruits should come within the very minimum of that period. But there are other cases where picking up to about a week before the show can be suitable, as mentioned above. There is a wide choice of containers for the packing of fruit, and there are several ideas to be followed. One easy way of carrying quite a lot of fruits – to enable several entries in a show to be made – is to line shoe boxes for use in holding soft fruits and to pack three or four of these into a larger cardboard box. This, for example, would be a possibility if travelling home from an allotment on public transport, on the assumption that the box can be carried on one's knee on, say, the bus. It does require a little thought to ensure that the excellent fruits that are picked look just as good when judged at the show, but it is very worthwhile and not by any means an impossibility.

Vegetables

The basic consideration of exhibiting fresh-looking produce applies just as much to vegetables as to fruit, and other parallel considerations are the retention of stalks and 'bloom'. The latter can be seen on well-grown peas, cucumbers and kohl-rabi, and it is most important to avoid finger marks, so handling by the tails or stalks is necessary. And, of course, 'picking' will have to be done with sharp scissors in order not to break the stalks on beans and peas. Furthermore, as soon as the vegetables are harvested they must be placed into suitable containers or onto sheets of paper rather than onto the ground. Tissue paper-lined trugs, baskets or boxes, or alternative containers with a soft lining, will therefore be required 'on site', and clearly for the exhibitor who grows vegetables on an allotment some distance from home and has to rely on public transport, an entry in numerous classes at a show will necessitate harvesting over several days. If you are making a debut, I would not advise doing so on such a massive scale, and, in fact, in Chapter 12, dealing with a selection of vegetables for exhibiting, I have deliberately chosen a limited number of kinds on which to concentrate for a start. There are, however, some vegetables which can benefit from being 'lifted' well ahead of the show date and some others that will not improve but can be kept a few days without deteriorating, provided that they are properly treated. Thus it is feasible to spread the task of harvesting over a period of about a week and make a substantial entry in the show.

The major difference between vegetables and flowers and fruit, accepting that they all need careful harvesting and handling, lies in the lifting of root crops and the requirement for washing them and also washing various others. Seemingly these tasks demand greater effort, but unnecessary strain can be avoided by some simple measures which will also prevent damage to the exhibits. A sensible plan for lifting long carrots and parsnips, for example, is to scrape away some soil from around the crowns and fill the depression with water until the soil is soaked thoroughly. If the soil is soaked to the depth of the roots, they can be eased out by a grasp of their tops, a steady pull and a slight twist. Any resistance should be met by more soaking or, if a fork or spade is used, it should be placed to one side so that the root is not bruised when levered out. Patience and care are needed. This practice of soaking the soil can be extended to other crops such as leeks.

Special care is called for when lifting potatoes, and a flat-tined fork of the maximum possible depth should be thrust down and away from the outside of the haulm (foliage) which should be grasped and pulled firmly, simultaneously with the weight being taken underneath on the fork. The tubers must then be separated carefully and placed onto paper or sacking, and not onto stony ground because the skins are very delicate. The next step is to choose those that have the potential for being exhibited because of their nice shape and size and absence of blemishes and so forth, and decide about washing them. Some exhibitors prefer to let the skins set for a few hours, whilst others like to remove the dirt straight away. A running tap or gentle stream from a hose should remove the bulk of the dirt from parsnips and other faily robust-rooted vegetables, but this could be too severe for tender-skinned potatoes; and a bowl of water and a gentle swill, with light use of a soft sponge, is to be preferred. Grains of dirt within the small cavities of the eyes will have to be gently probed by the lightest finger pressure, and not by a tool that might pierce the skin. The question of where and how the washing of vegetables is done will depend upon where they are grown and what facilities exist. Some allotments do not have taps and usually hoses are banned, but, of course, at home there might be justifiable complaints about the mess made in the kitchen! Whether washed or unwashed, however, the vegetables must be protected from damage whilst being carried – perhaps from allotment to home – and tender-skinned ones such as potatoes must be made snug in cloth-lined containers; and they must be covered to exclude light which would give an unwanted green hue to the skins. Should vegetables have to be collected from a muddy allotment in the rain, inevitably the container linings will become wet and dirty and will have to be replaced: a nuisance, of course, but better that than dispensing with the precautionary measures, and damaging potential winning exhibits, thus spoiling their chances of success. In fact, when working in conditions far from ideal, one is apt to be rather hasty, so the call for simple protective measures is even greater than might otherwise be the case.

Possibly the root vegetables take the most time to clean, and, owing to their tender skins, the potatoes more than others – although beetroot certainly require careful handling. After a thorough clean, some attention to detail is needed. This includes removing hairs or 'whiskers' from carrots, and this can be done by using a sponge and employing a circular motion whilst washing them. Small side-roots must be removed by deft use of finger and thumb. Similar steps should be taken for parsnips, and if any of these vegetables have 'fanged' – that is, divided, roots, they ought not to be considered for showing. Fortunately, beans and peas, among others, do not require to be washed, although any 'smuts' on broad beans should be wiped away. Peas, in fact, need to have their delicate covering of 'bloom' unmarked, so certainly no wiping or washing should be carried out. Cabbages must have all vestige of dirt between the leaves washed away under a tap or by a gentle stream from a hose, and similarly lettuces and celery, but extra care has to be taken otherwise broken leaves and stalks will result. It will be found with practice that holding them head downwards under a tap will enable most of the dirt to be removed without damage occurring. Leeks can also be dealt with in this way, but swilling in a bowl of water and a few judicious squirts from the tap within the foliage will help.

Thus it is that all vegetables must be clean and, after any necessary washing, will come the need for drying which can best be done by a flow of cool air. Dabbing with a soft cloth will probably be necessary, and all vegetables must be placed in a cool place whilst awaiting conveyance to the show, whether they have been washed or not. They should be lightly covered with paper or thin cloth, but do beware that newsprint does not transfer itself, and potatoes must be adequately protected from the light because, as previously mentioned, in no time at all they will commence to 'green' and judges always look for this fault. So when they are dry, they should be wrapped individually in paper. Of course, the time that vegetables have to spend in storage depends upon the arrangements for getting to the show, when they can conveniently be harvested, whether they need to be lifted a while beforehand to help them mature or whether, on the contrary, they rapidly lose their freshness when picked.

Taking first the vegetables that will look their best if harvested close to show day, we have all the members of the brassica family – Brussels sprouts, cabbages, cauliflowers and so forth – and soft-leaf kinds such as lettuces and spinach; also peas and beans and any kind with a sappy stem, rhubarb being a good example. Whilst some kinds will last for several days in a cool place – or you may prefer to put them in the refrigerator – they will show tell-tale signs of ageing: for example, the calyx on a tomato will shrivel and the blossom on the end of a cucumber will die, and these features should desirably look fresh when the exhibits are being judged.

Obviously the staying powers of individual kinds of vegetables would need to be assessed, and other vegetables can be harvested earlier. Top of the list come onions and shallots – they, almost alone, need lifting at least a week earlier if they are to be shown with a nicely ripened skin. About a fortnight before the show their roots need to be loosened from their hold in the soil by a gentle lift underneath with a fork and soil can be carefully scraped away to expose more of the base part of the bulb to the air. A week later they can be harvested, and laid out on a wire frame or onto sacking in a covered but airy place. If, on the other hand, the class to be entered is for 'Onions as grown', the tops of the onions, which then need to be retained, will rapidly lose their attractive greenness – in which case early harvesting is not to be recommended.

Apart from onions and shallots which usually benefit from early lifting, potatoes will come to little, if any, harm if dug up a week before a show – provided, as already stated, they are wrapped up to exclude all light. Beetroot will soften fairly quickly after lifting, turnips probably less so; in both cases, where necessary, they can be stored in a cool place for a few days without marked deterioration. Parsnips, too, will cope with one or two days out of the ground if after washing they are wrapped in a cloth which can be lightly sprayed with water. Although there is nothing to be gained apart from some time and convenience, these particular root vegetables do offer a little scope for harvesting at an earlier date than would be suitable for some other crops. But, if necessary, some of the other vegetables can be accommodating, provided that they are dug up with stem and root intact or cut with a long stem. Thus, cabbages can be stood in

25

buckets of water to keep their heads fresh and stems of globe artichokes can be placed in water and a little cut off the bottoms each day to ensure a good uptake. Cauliflowers, if hung upside down in a dark, cool place, should be reasonably fresh after a few days of such treatment, and some exhibitors, seeing that curds are fully formed, will follow this practice rather than allow the possibility of their losing points because the heads are over-developed by delayed cutting. By the same token, if a large collection of runner beans of uniform length is required, picking could be staggered over a few days, bearing in mind that on well-grown plants pods will develop rapidly. The pods must be kept fresh in a damp cloth or with stems in water. All these practices need some experience and, to begin with, I would choose the optimum method of ensuring fresh exhibits – which apart from onions and the like, and possibly potatoes, which need early lifting, means harvesting them on the day before the show or thereabouts.

Having established the point about freshness, which will be sought when selecting vegetables for the show, the other features to look for are uniformity in size and shape and good, matching colour. You can gear yourself up to this during the week before the show by, for example, strolling along the rows of beans or casting an appraising eye over the cucumbers and tomatoes in the greenhouse. This will give some idea of where you will be choosing the exhibition-quality specimens. At the time of harvesting it will have to be borne in mind that during the further period of some 15-24 hours up to judging time, colour will in certain cases, and particularly in the case of tomatoes, continue to develop. Any 'fruits' with a very ripe complexion at the time of picking should therefore be ignored. Of course, the length of carrots or parsnips is a hidden mystery, unlike, say, beans which are visible in their entirety; but girth can be judged when soil is scraped away to create the depressions for the soaking process mentioned earlier. 'Spares' will be needed to insure against the unlikely but not impossible chance that slight damage or flaw might be spotted at the show venue, but clearly in some cases a reserve exhibit is not a feasible proposition. It is difficult enough to produce a matching pair of marrows, and very few people would wish to take a spare pumpkin along to the show, for example; but on the other hand, a small blemish might not be noticed on a pea pod until staging time, and I would say that a sensible reserve of two pods for every ten for peas and beans, and perhaps a similar proportion for tomatoes and potatoes would be of about the order required. A reserve onion for each set of six would similarly be about right if that can be managed. Individual judgements will have to be made but, in general, I think it is but wise to have a small reserve of at least the items that are not so obviously bulky as to render the policy impracticable.

It will be seen that there are many similarities between fruit and vegetables so far as exhibiting is concerned, not least of which is the need to retain stalks which can be used as handles, thus enabling unseemly marks on skins to be avoided. Freshness and even size and matching colour also obtain, and although in the case of neither type of produce do the judges taste the exhibits, it must all look fit to eat if it is to win. And the immaculate appearance that is so desirable can only be achieved if a little care has been taken with harvesting. To take two examples: Brussels sprouts can be cut from the stem of the plant with a sharp knife so that they can be exhibited with a stalk neatly trimmed; and rhubarb sticks should be grasped firmly at ground level and pulled vertically. The alternative methods of snapping the sprouts and pulling the rhubarb by the top part of the stick might very well do for culinary purposes, but the risk of losing the stalks entirely of the former and the bottom inch or so of the latter is not worth taking. Care – or perhaps I should say 'caring' – is essential to successful showing; but so it is with many other pursuits where one seeks success. Surely it is all part of the involvement and not a chore.

5. Getting to the Show

Having safely harvested the produce, it must be conveyed to the show with equal care and then staged so that it presents the best possible image to the judges. There are obvious advantages in having a car or, better still, a large estate car, but many exhibitors manage extremely well by using public transport, and in the case of exhibiting at places like, say central London, parking difficulties can militate against using one's own transport. I have not witnessed recently what was once a common scene of exhibitors trundling their produce by wheelbarrow to the village green – although undoubtedly this can be a fairly safe method provided that tender items are sufficiently protected from damage. Nor latterly have I seen competitors carrying bunches of sweet peas wrapped in tissue paper down the lane to the show, but again this is a safe method wherever it may still seem apt. The usual scene that I observe is of lots of cars backed up towards side doors of halls or marquee entrances and of exhibitors with wives, husbands and friends – and sometimes aided by the whole family – working away steadily to transfer the contents into the staging area.

Flowers

Packing exhibits into a car does require patience, and whilst if passengers do not occupy all the interior, it seems an attractive idea to have containers such as plastic buckets partially filled with water standing in the 'well' between the seats, flower blooms so carried can receive quite a drubbing from repeated vibration against the back of seats or the door interiors. So this point needs consideration; none the less, I can think of few, if any, other methods of conveying mop-headed hydrangeas and blooms of that type that would guarantee less damage. All sorts of ingenious ideas are put into practice including simple ones like using plastic-covered mesh on the top of a bucket or deep tin to hold each stem of a flower or having a number of small-diameter vases fitted into a wooden stand to achieve the same purpose. Having the flowers continuously in water ensures freshness, but some

28

exhibitors prefer to carry their exhibits 'dry' bearing in mind that in water they will continue to develop and might reach a point of being slightly past their most perfect stage by judging time. Experience will help in assessing this, but if I were a novice exhibitor, I would place most emphasis on freshness and do everything to preserve it as a matter of priority. If, however, carrying flowers in water is ruled out for whatever reason, and flower boxes of the type used by florists are not available or not convenient, a similar arrangement to some of those used in the 'water method' can be employed. I have seen, for example, a deep cardboard box fitted with a piece of mesh towards the bottom and with another piece fitting the top or a lid suitably perforated with holes, so that flower stems can be securely held in place. The blooms must be kept free from resting on the top, and that is a matter of cutting the stems to a suitable length; and a piece of cotton wool placed under each bloom will give added protection (see Plate 4).

Flower boxes are the most commonly used type of containers for conveying blooms to a show, and they can be carried by car, on the train or when walking. If they are not to be laid flat 'in the boot' or on the back seat of the car, and have, by force of circumstance, to be stood on end – as might happen when travelling on the 'underground' – the flower stems must be firmly anchored to prevent movement. Various means of improvisation are possible; short pieces of cane wedged across the box just under, but clear of, the bloom head, and also towards the bottom of the stems, should prevent movement, and, of course, the flowers must nestle on something soft. The boxes are, therefore, usually lined with tissue paper, although the slightly more expensive method of using cotton wool is preferred by some exhibitors. An intriguing way of carrying daffodil stems is to pack a flower box lengthwise with 'rods' made from rolling newspaper very tightly to a diameter of about 2 cm ($\frac{3}{4}$ in) and then covering them with tissue paper. The flower stems are then trapped firmly but gently by slipping them within the 'rods' which are sufficiently pliable to make this possible. The flower heads rest on top of the tissue paper, and with a little practice, several rows of stems can be packed longitudinally in a box. An important feature is not to let a flower head come into contact with a wet stem, and stems must be wiped before being inserted into their positions in the box.

There was a time when cardboard was considered unsuitable because it tended to dry out flowers, and it was then customary to make up very large wooden boxes for transport by rail. I do not know of any exhibitors today who rely on such a method, and I am not aware of complaints about flower boxes which are usually obtainable from florists and are made in various sizes. There are, as with most things, some limiting factors, and these concern their length and depth: a delphinium spike will need both a deep box if the lid is to be fitted and one also with considerable length. Purpose-made wood boxes are to be seen around, but an alternative if using ordinary florist's boxes is to make holes in one end at an appropriate height to enable the stems to protrude by several inches. And, if sufficiently courageous, and the box method is thought unsuitable, the exhibitor will wrap the blooms gently in tissue paper and carry them by the stems, even laying them on a luggage rack in a train under a watchful eye.

The essence of it all is to prevent potential winning blooms getting spoiled in transit, and it is worth considering that many exhibitors take the view that less damage, if any at all, will be caused by blooms being in contact with each other than by being pressed against man-made material such as the side of a box. A method, in fact, that was considered successful in yester-years, but which I have not encountered, was to parcel flowers by first wrapping them in tissue paper and then rolling two or three bunches so wrapped into one or two sheets of brown paper. The belief behind this was that a light shake would free any compressed blooms and no residual damage would remain. I could not recommend this practice today, but I think that it demonstrates that it has always been accepted by growers and exhibitors that if a flower is to be in contact with another material thing, it had best be another flower, unless something softer can be substituted.

Fruit

Unlike flowers, fruits will inevitably bruise each other if pressed together in any way, so they must be carefully packed to ensure their separation and, at the same time, to prevent their rolling around in their containers. Ideally, fruits will be picked and placed in their cotton wool- or tissue paper-lined containers, kept like that in a cool place, and taken to the show without further handling, although as a matter of convenience, several small containers will need to be packed into one larger one. I have found that strong cardboard boxes – which abound in large grocery stores or supermarkets – are very useful for holding considerable quantities of various items, provided that they are sensibly packed. Jars of jam and other fairly solid items that are to be exhibited can provide the bottom layer, and with newspaper placed on top of that, containers of soft fruits can form a top layer which is covered with tissue paper taped to the box to prevent its blowing away. A sheet of thin polythene stretched over the top of the box, but not touching the fruits will be necesary in wet weather; and this should be anchored into position by some form of tape. It is by attention to such apparently small details, that damage can be avoided, say from an unexpected rainstorm while carrying the boxes out in the open. A further obvious point that many of us choose to ignore from time to time is that all the materials needed for packing purposes can be got together early in the week before show day. It is not conducive to domestic harmony to rush all over the house just prior to setting out for the show, looking for the missing polythene or whatever; and the odds are that in doing this something will be damaged!

Vegetables

As with fruit, damage from careless handling is easily avoidable by the exercise of some patience and care whilst packing. Some 'double handling' will be inevitable because of the size of some vegetables, such as winter cabbages – the hearts of which will often be formidably large. They will have been laying on a cold surface in an outside storeplace or sitting in a bucket with root intact and unless a large enough box is available, they may have to be carried as separate items. In any event, they must receive more than passing attention because, if well-grown, the leaves will be turgid and they will snap easily, and this will lose them points when judged. The schedule for the show should be looked at (again) to determine what it says about 'roots'. If, as is likely, it says that they should be trimmed to about 7.5 cm (3 in) in length, they are best cut off with a sharp knife before being conveyed to the show simply because the extra 30 cm (1 ft) of stem would be a nuisance in most circumstances. A soft cloth should, however, be used to protect the exhibit from damage during the journey, and in no way should it be treated less reverently than other exhibits simply because it is a cabbage!

Light will discolour not only potatoes, so carrots and parsnips and other root crops should be kept covered with a soft cloth. Rhubarb will need a damp cloth if the journey is to be of any length and the leaves will have to be trimmed before setting off in order to assist packing or save space; usually about 7.5 cm (3 in) of leaf blade is shown, but should you be exhibiting 'forced' rhubarb, which has that delightfully juicy-looking pink stalk and a small unopened light green leaf blade, no trimming other than a little tidying up of the bottom of the stalk would be done (Figure 5.1). In trimming generally, experience will indicate how and when this is best done. With some vegetables, a certain amount of freshness is preserved by leaving the foliage on the exhibits as long as possible; cauliflowers will appreciate the cool

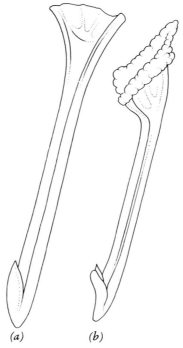

*Figure 5.1
(a) Rhubarb is
shown with the leaf
trimmed; (b)
'forced' rhubarb has
leaf intact*

(a)   *(b)*

wrapping of their leaves, which can offer some protection in transit, but a lot has to depend upon the packing arrangement that is feasible. Whether or not that applies, a major consideration is that there is a lot of work to be done on the staging table at the show, and anything that can be done at home without detriment to the produce will spare you from getting in too much of a flurry, and will be helpful to the organisers who are seldom able to provide abundant staging space.

Peas and beans nestling in their cushioned containers should present little problem, although, as with fruit, a plan will have to have devised for packing a number of smaller containers into a largish box. A really large pumpkin will possibly require the back seat in the car and its weight should prevent its tumbling off, but don't take anything for granted!

There is no doubt that in transporting exhibits to a flower show, the adage that 'where there's a will, there's a way' is apt. But it all depends upon exercising care. This is just as important with items for the handicraft section as for fragile flowers. The sort of experiences that I have undergone or witnessed are by no means unique, I am sure. I once, for example, had a pot of cactus in a stout box on the floor of an underground train and it was unfortunately kicked some distance by a boarding passenger. The box was strong enough, but I had not packed the pot firmly within so it ricocheted and tipped out the contents. The message became clear – make sure that pots and the like are immobile.

Equipment Needed for Staging

As well as the produce for exhibiting, a certain amount of equipment must be taken to the show to enable the exhibits to be properly displayed. This can conveniently be packed into a small case or bag, unless the show arrangements are that the competitors' own vases or bowls must be provided. If such is the case, I am afraid that that extra load has to be carried with you; it would help, however, to obtain purpose-made vases made of polypropelene because they will not require the precautions necessary with glass or china containers and will be much lighter. Needless to say, their very lightness could be disadvantageous, but if appropriate sizes are used for the taller displays, they should be stable when filled with water. Some sphagnum moss will be needed to pack flower stems in their vases, although it is becoming fashionable to use oasis or similar water-retaining material, even for sweet peas which are probably better set-up by having their stems enticed between thin reeds filling the vase. Whether one has a source of supply for particular aids will influence one's choice. A pen and a pencil are essential and the small name cards which will be of interest to the visitors to the show (in some shows they may be obligatory). Scissors and a knife should be taken along, a cloth and/or sponge plus a small, soft brush. Beyond these, some things are optional: 'onion rings' (see page 120) will in the sense that at most local shows they are not insisted upon enhance the appearance of a set of onions, and paper plates should be used for staging fruit unless plates are provided by the show organisers. Some vegetables are also commonly staged on plates – filled with sand for shallots – and baskets are also used. These latter are customarily filled with parsley to enhance the whole appearance of the exhibit, but the show schedule should be carefully studied to make sure if anything is ruled-out, or indeed if any special requirement for staging is stated, as is the case with shows staged by some societies, including the Royal Horticultural Society.

It always pays to stage neatly and attractively and to take along the essential materials. The ideal is to see what happens at National-standard shows, that is to say, shows at which standard forms of presentation are practised according to the rules of National Societies or the RHS. These may be held in London at the RHS Halls or elsewhere in the United Kingdom, and it is worth making a journey to them to see how exhibits are displayed on the show benches, and what items of equipment or other materials are required.

The exhibitor in Floral Art will naturally require some additional materials, including gutta percha, wires, pin-holders, oasis, pieces of driftwood, 'backcloths' and various other accessories, as well as vases, pots or bowls; and in the case of exhibiting in the Domestic section, items such as doillies will be needed. The importance of listing what is required and checking that it is all to hand when packing things for the journey to the show cannot be over-stressed. This applies not only to the accessories but also to the produce to be exhibited. It is far too easy to leave something behind as we all find when going on holiday – and a checklist is a 'must'.

6. Staging

Introduction

Hopefully the exhibitor will arrive in good time at the show venue and with all exhibits in good order. Unless the layout is familiar, it is wise first to have a look inside to ascertain where the staging area is and if, for example, several trips to the car need to be made in order to unload, where everything might be safely put. It can be disastrous not to do this methodically, because for the many people rushing about, boxes not stowed safely become obstacles which they fail to jump successfully – with damage to themselves and to the contents. I do not want to paint a picture of chaos, but it has to be accepted that the morning of a show is a busy time for everyone involved, and whereas at larger shows there is usually more space available for staging – and staging may be spread over a longer period of time – at local shows space is often at a premium. In a school, the corridors are sometimes pressed into service as useful staging areas which will accommodate several tables, but in the circumstances that just a large room or a marquee is available with no adjacent area, one section has to be used for staging. This involves the stewards or committee in clearing up and filling the space created with decorative displays or even with a raffle table. Obviously it is preferable to keep the business of staging away from the show hall itself, particularly if the hall is nicely carpeted or has a highly polished floor. It is impossible to avoid some mess from the damp newspaper, wet leaves or water spillage that so often occurs, and this is best kept to a place where a good clean-up can be carried out without interfering with the show.

The exhibitors can help a lot in 'keeping the place clean', and it is easier if they are not racing the clock and work methodically. So, having spied out the land and determined where to put your exhibits and other paraphernalia, it can all be taken to a staging table for unpacking and, in the case of flowers, for placing in water, assuming that they have not been conveyed in containers of water in the car. If exhibits for the Handicraft or Domestic sections are included in your entries, they can, of course, be placed in position on the show bench or table so that they at least are out of harm's way. No 'dressing' of such exhibits has to be done, other than their being neatly laid out, and there is no point in keeping them in the staging area. An exhibitor's card must be placed 'face downwards' (with the name of exhibitor concealed) by each of these and all other exhibits, and having made your entries to the show secretary in advance, the cards needed should be available on a committee table in the hall or marquee. If there is no queue at that table when you arrive, it would be best to collect the required cards then: but if the show secretary is being harassed, as sometimes happens, it would be of help to wait a while and collect the cards later. Where late entries are acceptable – that is, the rules do not insist upon their being made earlier than the day of the show – you may find yourself with several others waiting round the committee table whilst the entries are booked-in and cards made out. Patience might need to be exercised in such a situation.

Some exhibitors make a practice of booking spaces on the show benches by placing their cards at points of their choosing in the various divisions and classes in which entries are being made. This does of course let the show secretary or stewards see how the spaces are being taken up and it might help them to adjust the layout accordingly, but this would not always be the case bearing in mind that in a class without parameters – for example, 'A Vase of Hardy Herbaceous Perennials' – a card would give no indication of the grandeur of the exhibit concerned. And if there are several cards on the tables without accompanying exhibits, it is a matter of sheer guesswork where to place your own card if the intention is to secure a good position where your exhibit will not clash with adjacent ones in respect of colours and so forth. Therefore I would not advise other than to collect your cards at a convenient time and to place them with the exhibits as and when the latter are ready for displaying on the show benches. At that stage, of course, there might well be something to be gained by selecting what appears to you to be the best spot for highlighting the exhibit. I know that some exhibitors take pains to do this, and certainly there is little point in hiding one's exhibits in the back row or in placing them next to others which seem obviously to outshine them, but it has to be remembered that judges are very thorough in their task and are unlikely to be over-influenced by factors which do not relate specifically to the overall quality of the individual exhibits as they come to examine them each in turn.

In staging generally, so much depends upon the facilities provided at each show and your own arrangements for getting there. The 'professionals' staging trade exhibits and societies entering massive collections of vegetables will expect to 'burn the midnight oil', otherwise they could not complete their tasks. Only recently, I witnessed representatives of a nursery specialising in sweet peas selecting hundreds and hundreds of blooms and staging them into the night to produce a display that was impeccably beautiful: but they were not alone because also in the large marquee were a number of amateur exhibitors involved in setting up vases of delphiniums and mixed flowers and some floral artists intently building pedestal arrangements and other stunningly lovely exhibits. Plenty of space was available, and there was no need to hurry: all that was necessary was to stay awake! Similarly, one will find exhibitors at such venues as the RHS Halls working steadily into the late evening and again early the following morning unloading the Exhibits.

But let us assume that you are showing locally and have something of the order of three to four hours in which to do the staging; and that a spot has been found in the staging area in which to work. Having disposed of any handicraft or domestic section exhibits by placing them in position on the show bench, the flowers should be attended to by getting them into water if they have been brought dry to the show. If you are able to bring your own containers for this purpose, you need only to find the water supply point and then carefully unpack the flowers from their boxes and steep their stems, the bottoms of which if at all dry should have a small piece cut off – at a slant. If you have to rely upon using vases provided by a society or show sponsor, please do not take more than the absolute minimum number which will enable you to manage your staging. It is not uncommon to find at some shows that exhibitors have taken possession of lots of vases and having completed their task, have left behind on the staging table at least as many vases as the number they have actually used to contain their exhibits. This is awkward for show organisers who find themselves besieged with requests for more vases needed by exhibitors who arrive with all too little time to spare. Provided that some large vases can be obtained, just one or two can hold largish numbers of stems, particularly of roses and other fairly robust flowers which will not become damaged by light contact with each other; and on the whole, the attitude to adopt is to settle for something a little short of the absolute ideal, if other competitors are to have a fair share of the available facilities. But that does not mean to say that everything possible should not be done to ensure that flowers are exhibited in a completely fresh and undamaged state.

Plenty of experienced exhibitors prefer to select their

blooms from their boxes and place them straight into the vases that are to be displayed on the show bench, thus omitting the interim stage of placing them first in other containers of water. This can save some 'double-handling' and with a practised eye it becomes possible to take a stem from its box, hold the flower towards the light and make an instant judgement about its merits and a decision whether a 'reserve' bloom needs to be used. The process is helped by the labelling of individual blooms (by using named adhesive strips or, where possible – for example with daffodils – by writing on the lower part of the stem with a ballpoint pen), so that they can quickly be related to the particular classes in which they are being entered. But the experienced exhibitor also knows about the recovery powers of flowers and the time intervals between their being not quite open and fully open, this being very important in the case of stems carrying several blooms – such as sweet peas where the exhibitor wants the top bloom to be open and the bottom bloom still looking completely fresh. For the beginner, though, I would recommend getting the flowers into water as soon as possible, to ensure that they are kept as fresh as possible.

In giving prior attention to the flowers, fruit and vegetables should not be ignored. It might be possible to store them in their containers under the staging table; polythene coverings should be removed to prevent 'sweat-ing'. On a hot day the boot of a car would certainly not be a suitable place for them, and if there is room to do so, it might be possible to spray the green vegetables – cabbages, lettuces and the like – with a mist of water as they lay under the table in their boxes, the lids having been removed to permit circulation of air. Flowers, however, are likely to take relatively longer to arrange than fruit and vegetables, although a large-scale collection of the latter would be very demanding in time; and unless one is involved in amassing such an exhibit, I think it wise normally to attend first to staging the flower exhibits. Should you, in fact, be involved in the Floral Art section, then that being the most demanding in time so far as flowers are concerned, a start should be made there. This is likely to take up at least the first hour of precious time even if only one entry of some substance is made. It is difficult as a beginner to cope satisfactorily with entries in a number of classes all taking considerable time; if it is wished to make an impact on the scene with one's initial entries, the spread and diversity of classes must be kept to a sensible minimum.

The Staging of Cut Flowers

Thus, leaving aside the Floral Art and assuming that one is not making a dramatic start by staging a large group of vegetables requiring elaborate arrangement, let us deal with staging the flowers. The whole idea is to enhance their attractiveness without losing their natural beauty, and normally 'own foliage' is all that is required. Indeed additional foliage can disqualify an exhibit, and the rules in the pertinent schedule or Handbook must be clearly understood. In some cases – for example, roses – the foliage would be expected to be growing on the stem, but this is not always possible in certain flowers, as in the case of stems and leaves growing from a bulb, and then it is customary to insert a few leaves behind the stem to improve upon the otherwise bare look. Where foliage is inserted in the vase separately from the flower stems, it must be fresh and clean otherwise its very object of enhancing the appearance of the exhibit will be lost; and if shown on the stems, its condition will be taken into consideration by the judges in pointing the exhibit or in making an overall assessment of its merit. So, it is most important to be as attentive with the foliage as with the blooms. May I stress, however, that the rules concerning foliage must be followed, and if there is any doubt about what is allowable, a word with the show secretary beforehand is advisable. Having done that or not, at the time of staging it is imperative that the show schedule is at hand and that the specification for each class in which an entry is made is re-read. It takes but a few moments, but it can save a lot of disappointment and embarrassment.

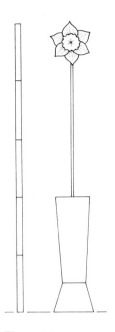

Figure 6.1
Where feasible,
length of flower
spike should be 1½
times height of vase

In making the exhibit look attractive, the proportion of stem length to that of the vase is important, and where possible the stem should be long enough to stand above the top of the vase by some one and a half times the depth (Figure 6.1). Stems will have been cut as long as possible, but should they appear to be shorter than desirable, the only, but useful ploy is to insert them in the reeds, moss or oasis packing the vase, to a minimum depth consistent with their not toppling about. It is not desirable to give the impression to the judges that the stems are only just held in position and, therefore, in the exhibitor's mind, rather too short: but equally one does not have to plunge the stems deeply into the vase as a matter of course. The stature of the vase should be no more than consistent with that of the flowers, an observation that may seem a matter of common-sense: yet it often happens that some nice-looking blooms are staged with their heads barely poking above the top of the vase, giving a flat and uninteresting appearance to an otherwise good exhibit. Another seemingly obvious point is that stems have to be inserted one at a time, and if moss is being used, it can be tucked round each stem to hold it in position, without, of course, jamming so much moss into the vase that it makes the insertion of further stems very difficult. If moss is scarce, but a little of it is available, paper can be used to hold stems in position and the moss can then be employed to make the top of the vase look more attractive than if it were filled with paper. No filling material should protrude above the top of a vase and some thought should be given to the plight of show organisers if vases are filled tightly with paper so that when being emptied after the show it becomes necessary to spend a lot of time in digging the paper out. Often exhibitors find themselves unable to be around to help in this task, and it therefore seems only fair if they bear these points in mind, although understandably they wish most of all to ensure the successful staging of their exhibits.

If it is possible to garner knowledge concerning some of the rather special customs of staging associated with shows sponsored by National Societies, so much to the good. Delphiniums, for example, are shown with the base of the spike of florets between 10-15 cm (4-6 in) above the top of the vase; that is, that amount of bare stem is regarded as all that is necessary to produce an attractive exhibit, and therefore the stem has to be trimmed if necessary to a length that permits it to stand firmly and show those few inches to the judges. Pieces of stem can be used to pack the spikes into position, and in order to create a pleasing appearance, two or three leaves can be inserted in the top of the vase, thus covering the packing material. A similar idea is practised by sweet pea exhibitors who insert twin leaves to 'finish off' the exhibit in a modestly workmanlike and attractive way (see Figure 10.4). These small quantities of foliage can be added at the conclusion of staging the stems, but in the case of daffodils for which it is usual to add two leaves to embellish an individual bloom and four or five to an exhibit of three stems, it is probably best to slip the leaves in first so that they can be firmed into position as the stems are placed in position in front of them.

Figure 6.2
Seven blooms: (a)
staged neatly in
rows or arcs; (b)
avoid haphazard
arrangement

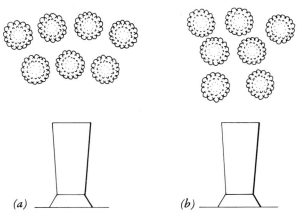

(a)   (b)

When arranging flowers, it should be borne in mind that only simplicity is called for, and elaborate and complicated exhibits are not likely to impress the judges. Taking as examples the staging of seven or nine stems, these can be displayed in two rows of four and three and five and four, respectively, with the blooms of the front rows fitting neatly below and between the blooms at the top (Figure 6.2): the stems of the flowers in the lower rows will therefore need to be shortened suitably. Large gaps are not wanted, nor should blooms touch each other. It cannot but help if it is seen that an effort has been made to present a neat arrangement that not only sets the blooms nicely but also helps the judges to see their merits, although, of course, it is their duty to study all exhibits carefully including any that are not carefully arranged. There are some customary practices that may be followed, and in point of fact these do not sometimes make judging at all easy. Sweet Williams are often shown in an all-round style, and if it is felt that this is what is expected within a particular society's ideas on the matter, then by all means follow that tradition. In general, however, I think it is best if blooms 'face the front' and are set up in rows of sensible numbers so that each bloom is easily viewed. It is imperative to check what the schedule requires by way of number of stems, and if the choice is left to the exhibitor, the temptation to go for large numbers at the expense of overall quality must be resisted: seven really good blooms all matching each other in their individual characteristics will beat an exhibit of nine blooms comprised of eight 'stunners' plus one bad one. It is surprising how often one sees good exhibits spoiled by the inclusion of one or two stems carrying dead or dying blooms, and this is notably the case where a vase of mixed flowers is called for. More often than not, no number of stems is called for, but there may be a stipulation that at least three or more kinds of flowers are to be shown, and sometimes this excludes some if not all of the kinds for which a separate class is provided in the schedule. If it is possible to meet the requirements with good quality flowers of the minimum number of kinds, it can only mar the exhibit if further kinds are included and the quality of those stems is inferior. Nor, where the schedule permits additional foliage, will it help to increase the girth of the exhibit with it to such an extent that the foliage dominates the blooms. So, 'padding out' must be guarded against.

Sometimes an all-round arrangement will be asked for, and the best procedure then is to select a fine centre-piece such as a multi-bloomed stem of Regal lily and work round it, placing each stem of flowers at a suitable height so that the blooms are nicely spaced and not crowded together. Turning the vase as you work will help to produce a balanced appearance. For a frontal arrangement flowers should similarly be nicely spaced and fanned out to make a widish spread – but watch for any stipulation in the schedule about overall measurements. The quality of blooms will be the main factor to be considered in judging, but the arranging skill will count in close competition and in some shows points are specifically awarded for it.

Figure 6.3
Bowl of pansies,
blooms well-spaced

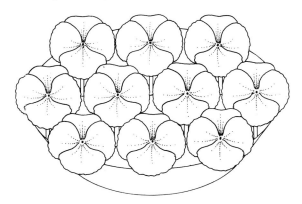

For some classes, a bowl or a 'container' will be requested. The latter covers everything that will hold water, but a bowl should always be a vessel with a width of mouth at least as great as the height: conversely, the height of a vase must exceed its width. A small bowl is usually the container chosen for staging pansies, but the fact that it is round should not make it incumbent on the exhibitor to set out the blooms in a circle. Far better to have them set in rows across, perhaps with three, four and three blooms in each row from the rear to the front of the bowl, to make up a total of ten if that is the number asked for (Figure 6.3) Oasis is used in many cases instead of damp sand to hold the stems – which should be long enough to lift the blooms clear of whichever packing material(s) is used. The bowl should be just large enough to permit the blooms to be spaced so that they do not touch each other and that also the very minimum of the packing material is seen.

'Dressing'

In watching competitors prepare their blooms for staging at national shows and at shows of similar standing, one becomes very much aware that a process akin to that of the make-up artist is considered to be essential in order that the chances of winning are maximised; except, of course, no touching-up with paint and the like is allowed! Petals are combed and teased into position by dexterous use of a camel-hair or sable brush and stems are straightened by gentle pressure of warm hands – I mention one or two other points in this connection in Chapter 10 on the choice of flowers for showing: and all in all, it seems sensible enough to take steps not so much to hide faults but to turn them into virtues. But whilst the beginners would be well advised to remove all spots of dirt or dust with the tip of a soft brush, I doubt if, until experience is gained, some of the dressing measures that abound would do other than eat into the precious time that could best be spent on the other aspects of staging. This assumes, of course, that everything will be done to stage clean and fresh flowers and foliage.

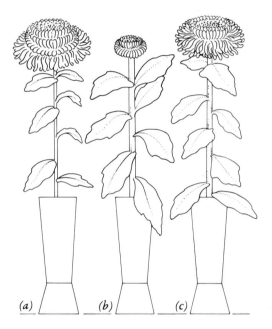

Figure 6.4
Bloom should be in scale with stem. (a) Stem too thin; (b) bloom too small; (c) correct

(a) (b) (c)

Summing up regarding exhibiting cut flowers, the aim must be to present for the judges' approval an exhibit that has obvious merit and no obvious faults. The latter cannot be hidden but they can be played down by attention to detail during staging: the presence of dirt or insects should be regarded as inexcusable and yellowing leaves denoting ageing should be eliminated. Stems should look to be in proportion to containers and indeed to the blooms which they carry: a tall thin stem carrying a heavy-looking bloom or a very thick one bearing a number of dainty blooms should be avoided (Figure 6.4). And the arrangement should immediately please, it being held theoretically that first impressions last longest. This may not be an infallible rule, but I think it can be said with some certainty that a bad first impression takes a lot of overcoming, and it really would be expecting a lot of a judge to find sufficient time to seek out any virtues of an exhibit that may contain several faults. In complying with the schedule, there are no 'ifs and buts': the number of stems, spikes, blooms or of whatever stated therein must tally with the number exhibited. And any stipulation about foliage must be carefully followed: if, as an example, the schedule states that a bloom must be 'with leaf or leaves' – as is stated in camellia competitions – then at least one leaf must adjoin the stalk carrying the bloom, otherwise the exhibit has to be disqualified. It will soon become second nature to comply with the requirements, and there is no more difficulty than that of a little self-discipline; it is very satisfying to feel that one is doing things 'professionally'.

The Staging of Fruit

The aim is the same as for flowers or for all other sections of the show: to present an attractive exhibit without frills, and there are certain traditional ways of achieving this which are best followed even if the schedule does not make them mandatory. Starting with grapes – which I admit are not by any stretch of imagination the most commonly shown fruit – we find that a rather special method of showing them on stands has been used over the years. It is also a mandatory stipulation in the RHS Handbook, so if the local show schedule states that RHS rules will apply, it is strictly required that the stand method be used. In any case, it is a very attractive way of displaying bunches of grapes and it saves them from damage on the show bench. The stand is easily made from a piece of thin plywood about 32 × 20 cm (13 × 8 in) with a bracketed piece of wood at the back to hold the stand at a slight slope in the manner of a photograph frame standing on a table. It is faced with a layer of cotton wool which in turn is covered with white tissue paper, secured by drawing pins to the back of the board. A small metal hook is fixed at the back of the board and overhanging the top so that a piece of raffia tied to the 'handle' of the bunch of grapes can be suspended in a natural-looking way. Black grapes look particularly attractive hanging against the white backboard.

The RHS Handbook contains much information on the subject of staging fruit, and a number of special points are made. Apples should be shown eye-uppermost, and a good polish on one's sleeve is taboo: all fruit, in fact, should be shown with its 'bloom' unspoiled. Except in the case of grapes and possibly large water-melons and suchlike fruits of great substance, fruit is best exhibited on plates. White china is provided at some shows, but an alternative for local use is the 'cardboard' plate. The china plates are customarily covered with white tissue paper, and a doily in the centre of cardboard ones will prevent fruits rolling around; or a few leaves can serve the same purpose, but these must not dominate the fruits. A melon can be seated on vine leaves covering the plate to create a pleasing exhibit, and discreet use in this way of a few leaves will enhance the show: but it should not be expected to add to the points value of the exhibit, and the schedule should be checked for any mention of what is, or is not, admissible. I must confess, however, that I have not come upon any local rules banning the use of leaves for decorating fruit exhibits.

It is not possible to sit apples comfortably on long stalks, and a little wad of tissue paper will overcome the problem; and a small mound will enable you to sit one of an exhibit of six fruits in the centre of the other five, thus producing a nicely rounded effect to please the eye of both judge and spectator (see Plate 17). Pears are not stood on their stalks, and they can be comfortably placed on the plate in a circular outline, stalks towards the centre. Sometimes they are stood against a small column of tissue paper placed in the centre of the plate and sometimes one fruit will be placed sideways across the top of the column so that it lies slightly above the other four or five fruits as a sort of capping to the exhibit. The important thing, however, is to avoid over-elaboration and to try to present the fruits so that they are all visible at a glance and do not hide their light under the proverbial bushel.

Nectarines and peaches can be seated on wads of tissue paper as can very large plums, and smaller fruits – including all soft fruits – can be placed in parallel lines across the plate, with their stalks pointing away from the front of the bench

Figure 6.5 Dishes of goose-berries. (a) Staged in neat rows; (b) staged in haste, producing a hap-hazard effect. Note the broken stalk

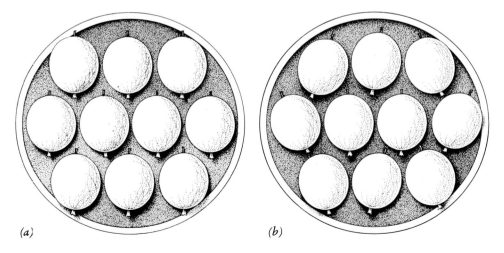

(a)   *(b)*

(Figure 6.5). It is also fashionable to arrange strigs of currants like the spokes of a wheel, the handles of the stalks meeting in the centre. And another practice is to lay berries on the plate in such a way that represents their shape: a tapering fruit would therefore require that the rows of berries across the plate were of diminishing size towards the bottom. Whatever method is chosen (see Plate 15) it must not create a fussy-looking exhibit, and it is no bad thing to conform with local practice of experienced exhibitors.

Handling of fruits must be by their stalks as much as possible: if, however, you are confronted with the problem that the weight of a large ripe pear could tear it away from the stalk, no chances should be taken, and the weight of the fruit should be taken by holding gently underneath either with two finger-tips or the palm of your hand preferably cotton-gloved. One reads also that plums should be held gently round their girth rather than by the stalk – presumably because ripe, fat plums will part easily – but I fear that without practice this could lead to marking of the 'bloom'. Perhaps a combination of a gentle lifting by holding the stem but taking the weight with finger-tips under the fruit would be best: I can only recommend that some experimentation is carried out with some fruits that are not needed for exhibition. As with everything else, practice will help you to arrive at the most advantageous method.

Staging fruits for the show bench demands attention to detail, but it could not be described as at all complicated or nerve-wracking. On the contrary, fruits are fairly willing 'victims' and do not resist strenuously – as do some flowers – the efforts of the exhibitor to get them to conform to his or her wish. So it should not be difficult to present exhibits that look neat and attractive, and really little more than that is required, always accepting that the fruits are of good quality.

The Staging of Vegetables

Whereas fruits would be expected to be clean at the time of picking, the harvesting of most vegetables has to be followed by thorough washing, and a fair amount of trimming of roots and foliage is also required. In the main, most of this will be done before setting out for the show, but as already mentioned, a degree of protection and freshness could be obtained by retaining leaves of cauliflowers during the journey to the show. As the vegetables are unpacked they should be inspected for any specks of dirt that may have been missed earlier on, and a soft brush or dampened cloth or sponge will have to be brought into use if needs be. Outside leaves of cabbages or lettuces that somehow have been damaged in transit will have to be removed. Although judges will note the sign of removal of leaves as an indication that all is not perfect, they will consider that an exhibitor who has submitted an untidy-looking entry has not done everything that he might have done. It is unfortunate, perhaps, that no credit can be given for an assumed accident; the judges can only judge as they find, and whilst they might speculate to themselves about what may have happened to an exhibit that looks a little bedraggled, they cannot make an allowance in compensation. It is simply the exhibitor's hard luck if an unforeseen accident to any produce has occurred: but at least a little tidying-up can be carried out.

'Tidying-up' is perhaps the essence of what is required, particularly with those vegetables in the brassica group. The wholesale removal of leaves is quite wrong, yet it is often perpetrated in innocence, in the belief that it improves the exhibit. It is true that leaves showing the ravages of caterpillars or the passage of time are undesirable, but stripping them off can but prove that something is missing, and an exhibit considered to be 'whole' is likely to be given preference. So a balance has to be maintained between removing broken foliage in order to show that you really care about presenting a tidy exhibit and retaining leaves with a few blemishes. Brussels sprouts are often abused by over-stripping, obviously in the mistaken belief that this improves their appearance. Perhaps it does from a culinary point of view and certainly I have seen them displayed as if they were ready to be cooked: but the pale green 'button' that is disclosed by peeling off several layers of its jacket of leaves will have little attraction for horticultural judges. So remove but very few if they look tarnished to the detriment to the exhibit. Cauliflowers are an exception in the brassica range, and before they are placed on the show bench, the leaves

Figure 6.6
Dishes of beetroot.
(a) Showing a broken root caused by careless harvesting and uneven trimming of leaves; (b) dressed and staged with care

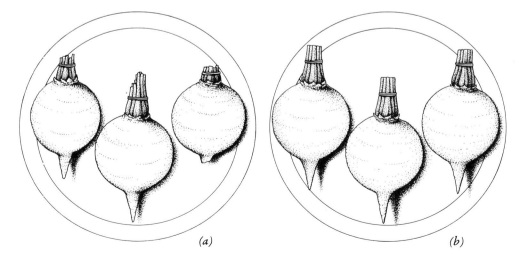

(a)   (b)

should be trimmed back so that the top of the cut stalks lies just below the crown. The idea is to show the maximum diameter of the curd.

In trimming foliage to an acceptable length, the show schedule must be re-studied to determine what is required, and it is usually more convenient to do this trimming at home. The normal practice is to reduce the leaf-stalk of beetroot, carrots, parsnips and turnips and similar kinds of root crops to about 7.5-10 cm (3-4 in), and this should be followed unless the schedule asks for something different (Figure 6.6). If it says '7.5 cm' (3 in) without further qualification, I think the exhibitor should comply literally, and this would entail measuring carefully and cutting the leaf-stalk to a length of precisely 7.5 cm (3 in). Pedantic this might be, and fortunately the word 'about' usually appears in descriptions of the amount of leaf or stalk to be shown, but following the demands of the show schedule is all-important if the exhibitor is to avoid disqualification. Certainly onions to be shown as ripe and not 'shown as grown' should have their tops removed to leave about 9 cm ($3\frac{1}{2}$ in) that can be folded down towards the neck of the bulb and then neatly bound with raffia or fine twine (Figure 6.7). Shallots can have their tops neatly bound in similar fashion – I do know that it is possible to devise a method of taking them to the show in their plate of sand without spilling the contents, but a jolt in the car can so often disturb the arrangement, that I think that this job can conveniently be done in the staging area. The bulbs should be all of the same size – in the case of pickling shallots, not in excess of 2.5 cm (1 in) in diameter unless otherwise stated – but if they vary slightly, it is worthwhile placing them so that the discrepancy in size is displayed to minimum effect. Remember that any ploy that is not cheating is permissible and it is reasonable to take measures

Figure 6.7
'Dressing' an onion bulb

to give the best optical effect rather than highlight defects by bad juxtaposition. A slight turn of a bulb here and there and adjustment of adjacent partners can be helpful. The bulbs can be stood upright in fine sand filling the centre of the plate – a 'cardboard' one will be acceptable – but a slight mounding effect can improve the appearance of the exhibit. It should not be too difficult to stage the bulbs in a symmetrical pattern within the circular outline of the plate if they are placed in position one at a time, working from the centre; but it is very easy to get the numbers wrong and end up with, say, thirteen instead of twelve bulbs, unless the staging is done methodically and calmly. A check on the number in the schedule and a careful count are essential. After that, the plate can be carried carefully into the main show area and placed on the bench together with its name card; and your exhibitor's card with the particulars of your name, address, etc. will be placed just underneath 'face down'.

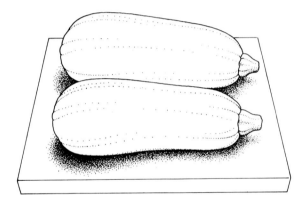

Figure 6.8
A pair of marrows,
neatly staged

Onions can be placed into position on their small staging rings, grouped in a neat arrangement. Plastic rings are available, or if a cardboard tube has been used, the rings can be cut to slightly different lengths. The arrangement can then slope gently from the back to the front of the bench, which helps to show each onion clearly. A group of six bulbs could be staged in a triangular arrangement, pointing towards the front; that is in rows of three, two and one bulb. The majority of other kinds of vegetables are commonly laid directly onto the bench, and this needs to be done neatly. Beans should be placed in a straight row, close to each other but not quite touching, and with stalks facing away from the front. A pair of marrows seem to present themselves better if they are placed close together horizontally to the front and not pointing towards it (Figure 6.8) whereas lettuces, like the beans, would usually be seen facing the front. But the use of a plate can improve the look of a number of vegetables including peas and beans, and certainly this is so in the case of potatoes which should be spaced evenly, with the rose-end of long cultivars outwards from the centre (Figure 6.9). Instead of being in rows, peas are sometimes placed on plates so that they radiate around it, stalks to the centre; and in collections of vegetables shown in major competitions, they are sometimes placed against a small block of hessian-covered wood, the long slightly curved pods flowing down towards the bench (see Plate 7).

Figure 6.9
Potatoes staged rose-end outwards

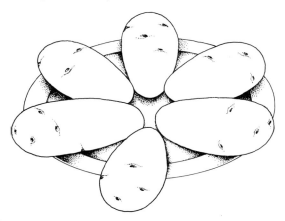

This sort of arrangement enters into the artist's world, and there really is a classical beauty about some of the mass displays of vegetables (see Plate 6) that are to be seen at major events. Of course, the staging apparatus is fairly elaborate from the beginner's point of view, although it can be easily constructed by a handyman (or woman). For example, a wooden frame similar to the snooker frame used to house the red balls, but with a fitted back, is used to stage cauliflowers. The curds are fixed within the frame and to the backboard by protruding nails which impale the shortened stems; and the frame can then be stood in an upright position. Sticks of celery and leeks are also secured so that they can stand upright at the back of a display, so suitable backboards need to be made for the purpose. Wicker baskets are used to house potatoes (see Plate 8) and the tubers are garnished with parsley which receives a special mention in the RHS Handbook as being admissible for the purpose when collections of vegetables are shown. It does not, however, collect extra points, although in adding to the attractiveness of the exhibit it could help towards a higher mark where points are specially awarded for 'Arrangement'. It is perhaps anomalous that whereas normally the judging of vegetables involves scrutiny of individual specimens, it clearly is not possible for this to be done if the arrangement of an exhibit – as is the case with large collections staged in major competitions – is such that it is not practicable for the judges to handle some of the items. To attempt to do so would involve dismantling the arrangement to a certain extent, and the public would lose the chance of seeing an immaculate exhibit: so I will not dwell on the thought that there must always be a chance that the splendid-looking leeks standing like guardsmen at the back of the arrangement could have a few hidden blemishes!

In any case, to begin with, I think the new exhibitor would be advised to concentrate on quality of produce, with staging being neat and simple; and this is likely to be in harmony with the local requirements. Should it be possible to produce adequate numbers of vegetables of good quality needed for a 'collection', I would advise against splitting-up kinds unless this can be done in such a way that the judges can quickly establish that the numbers tally with schedule requirements. For example, twelve runner bean pods could be divided into two groups of six, each group being placed to flank a plate of potatoes or tomatoes. This would produce symmetry, but it is not a practice that I would strongly recommend, and certainly nothing more complicated should be contemplated. If, for example, there is a class for a collection of four kinds of vegetables, a reasonable type of arrangement could be to make a centre-piece of six large onions, with a plate of tomatoes in front to the right and a plate of potatoes to the left; and two hefty cabbages could sit to the rear. The bulk has thus been evenly distributed from the rear towards the front of the bench, and not only does the arrangement look attractive, but it also lends itself to straightforward judging.

The question of numbers to be staged deserves special mention. Should the schedule simply state that numbers must be in accordance with the RHS Show Handbook, the only snag then is that for some kinds of vegetables the numbers requested are less if to be shown as single dishes in separate classes than if shown as part of a collection. A slightly irritating point perhaps, and one that might be overcome in a future re-write of the book. Very often, local show schedules have a table at the beginning or end stating the numbers to be shown, or they might state that numbers in a collection must be as for the separate classes of the appropriate vegetables. If under those circumstances, an exhibitor has a vegetable for which there is no separate class, it must be established with the show secretary what is required. Possibly the RHS Book would be the guide, and particularly so if the local schedule states that RHS rules will apply.

Having placed exhibits on the show bench, some, if not

all, of them should be covered with paper to prevent the light spoiling the natural colour. Potatoes will rapidly lose their pale cream or white colouring in favour of a subtle shade of green which is anathema to judges; parsnips will turn from creamy-white to a dirty amber shade and carrots will lose some of their glowing red. So covering them until close on judging time will be worthwhile, but try not to interfere with neighbouring exhibits either when placing the paper in position or when taking it off. I have seen some pretty clumsy exhibitors make rather a nuisance of themselves in disarranging other exhibits by careless or thoughtless actions, and it is unforgivable and unnecessary. Usually, however, there is complete co-operation from exhibitors in these matters, and furthermore the stewards will usually agree to remove any paper that has been placed in position by an exhibitor who is unable to get back to the show to do so; but the latter arrangement should not be resorted to except in an 'emergency'.

Everything then has been done to present the vegetable exhibits in their optimum condition: they have been carefully harvested and cleaned, adequately trimmed and gently handled: and they have been staged in a neat and orderly manner. If they do not win, they have to be beaten by exhibits of even better quality; and it will not be because of any failure to do everything possible to make yours the winning entry. Success is bound to come by proceeding on that basis.

7. Flower Arrangement

Flower Arranging, or Floral Art as it may be called, is universally admired, and one has only to go to the Chelsea Show to see that of all the superb features to be admired there, the displays of the floral artists draw the keenest attention. Of course, artistry with flowers has a homely connection, and there will be many who are involved in providing this touch of warmth with flowers in their church. So the art can be practised outside of flower shows, and can also be learnt and practised at Adult Education classes, or at various centres where courses are held throughout the year. Therefore it should be expected that the Flower Arrangement section of a show should draw plenty of entries, and more often than not this will be so. From the spectators' point of view, and also of help to the exhibitor, is the judges' practice of writing comments on the exhibitors' cards. They invariably do this as a matter of course (this is not the case in other sections of the show), and the time spent by the judges in doing this is far from wasted.

Basic Rules Regarding the Schedule

There is usually a wide range of classes in the Flower Arrangement section, covering elaborate exhibits based on a theme – such as 'A Film Title', 'A Summer's Day' or 'A Table Setting for an Anniversary' – and also the more simplistic designs using perhaps no more than three or five blooms in a modern style of arrangement; and usually there are one or more classes for petite exhibits. These last-named exhibits are not easier to arrange than large ones, particularly for those who have more thumbs than fingers, but at least the material required is not difficult to transport to the show! Whatever the class, however, great care must be exercised in reading the schedule, and an understanding of the general rules that obtain is, to say the least, helpful. They are usually based on guidance given by NAFAS (The National Association of Flower Arrangement Societies of Great Britain), or in simple terms the local show schedule might say that 'NAFAS rules will apply'. This is not so forbidding as might seem, and a grasp of what is required is very quickly gained. It is wise, however, to bear in mind that it is always necessary to comply with the schedule written for the show in which you are entering; it is also pertinent that the NAFAS rules for disqualifying exhibits state the first reasons as 'Including any component specifically forbidden in the show schedule or Omitting any component specifically required by it'. NAFAS recommend, for example, the use of a general term 'plant material' rather than specific descriptions such as foliage, fruit, grasses and so forth; but if the local schedule asks for 'Flowers and Foliage' and states nothing else, then to include fruits (which include berries) would lead to disqualification.

Having made sure that you have the correct components for the exhibit concerned, it is then necessary to conform with any dimensional limits that are stated in the schedule. These may apply to the exhibit itself or to the space or niche that is provided for it on the show bench. A usual layout at shows provides a series of niches made from wood or corrugated cardboard, and the schedule will normally state their dimensions. Obviously it would make for difficulties if an exhibit were too large to fit comfortably into its niche, but apart from that, it would mean disqualification, as would exceeding, for example, a maximum height, breadth or depth measurement, regardless of whether it were to be staged in a niche or not. Sometimes the word 'overall' appears against the stated dimensions, and this must be taken to include the three dimensions of width, depth and height, each of which must be within the stated overall measurement.

Strictly, and in compliance with NAFAS guidance, the cut ends of any fresh plant material must be in water or water-retaining material (oasis, for example), although there are exceptions made for certain items where the plants are either stemless or of a succulent or non-wilting nature. These include grass turf, moss, succulents (including cacti), lichen, and fruit and vegetables. So, a tell-tale sign of flagging of flowers or foliage would lead to disqualification where judging is based on the NAFAS Manual; but it would also be a matter of shame to see one's artistry spoiled by a lack of care.

Artificial plant material is inadmissible unless specifically allowed by the schedule; a matter I think of common-sense, and one really ought not to make an error in that direction. But perhaps the question of 'Accessories' is more difficult, unless the show is run or judged according to NAFAS standards – which permit accessories to be used unless specifically forbidden by the schedule. The one case in which there would be no difficulty is when the schedule states that 'accessories are allowed', because whether they are then included or not would not 'breach the rules': but it is seldom

Plate 1 A massed exhibit of apples

Plate 2 A 'corner' of Chelsea Flower Show, London

Plate 3 A composite exhibit produced from bulbs supplied by a local society

Plate 4 The use of flower boxes for begonias

Plate 5 The staging of gladiolus requires concentration

Plate 6 A collection of vegetables

Plate 7 Peas on a hessian-covered block

Plate 8 Potatoes in a basket

Plate 9 Exhibits of three vases, six stems of roses in each vase

Plate 10 The spacing of these 18 stems is most pleasing

Plate 11 Three fairly-well-matched delphinium spikes

Plate 12 A vase of mixed flowers

Plate 13 A fine fuchsia

Plate 14 Beetroot, two dishes

Plate 15 Redcurrants, three dishes

Plate 16 Blackberries, showing different methods of staging

Plate 17 Apples, two dishes

Plate 18 Shallots, with runner beans to the right

Plate 19 A flower arrangement of anemones and freesias

Plate 20 The iris show

that this wording appears, and I can only counsel that if you are in doubt, the show secretary should be consulted.

Basic Guidelines for Creating an Exhibit (Plate 19)

So much for the basic rules regarding conformity with the schedule. There are also basic guidelines for creating an exhibit. These can be learnt at the various classes held everywhere, or they can be self-taught by studying any of the many excellent books published on the subject. One is, of course free to experiment, and freedom of expression is something that can be practised with flowers just as much as with paints or pastels. An apt summary of what is required of the end result comes to my mind, and it is as follows:

'Unity, Scale and Accent
Balance, Rhythm and Harmony'

Taking the first requirement of 'Unity', it entails producing *Unity* an exhibit that obviously looks to be an entity and not several disjointed parts. The container has to blend with the plant material: but not necessarily match it in colour, of course! And where a composite exhibit is staged, perhaps by using more than one container and figurines and other accessories, all the components must form a co-ordinated whole and not appear as a number of separate units. They must, in other words, complement each other and have a distinct look of 'togetherness'; and I very much doubt if this can be achieved without some pre-planning before show day, so that at least the range of flowers and foliage and the right sort of containers and accessories are well-considered and are known to be obtainable.

'Scale' sounds like part of a draughtsman's vocabulary, and *Scale* really, I suppose, what is involved is akin to designing or 'drawing to scale', but whether you need preliminary sketches or not will depend upon individual abilities. Some can do it all 'by eye', but others may prefer to copy from a plan. Whatever method is followed, everything about the finished result has to be seen 'in scale'. A chunky china bowl would therefore need some robust-looking flowers, and a delicate crystal vase would be more suitable for flowers that do not look so heavily-built, albeit if the vase is of suitable height, long stems would be in order. Getting the right scaling can be a little difficult when composing a miniature exhibit which is either stipulated as of certain dimensions or might be something that is popular in some shows and appears year after year as a regular feature. An example is 'An Exhibit in a Thimble': this involves producing an attractive exhibit of tiny flowers and foliage with the appearance of a scaled-down version of a more usual-sized arrangement. It is easy enough to fill the small container with one or two blooms that are considered to be of a suitable size, but unless tiny flowers are used, it will not be possible to produce an arrangement that has a goodly variety, which is to be desired. Blooms of lobelia and virginian stock and thin strips of lavender leaves can be useful and often flowers of small stature can be found on the 'rockery' or in a sink garden. Where possible, individual florets with a suitable piece of stem should be used rather than clusters, the overall size of the heads of which is likely to look out of scale with the container.

Accent

'Accent' is achievable by placing blooms of a slightly contrasting colour or texture at strategic points which will draw the eye smoothly and easily to them. The effect must not be jarring or violent, and care must be taken not to turn a highlight into a glaring spotlight. Gerberas, which are expensive to buy and therefore need to be used sparingly, can be very suitable as points of accent, provided that the colour is chosen so that the blooms are not over-dominant: this should not be too difficult because they are available in pastel shades as well as brighter colours.

Balance

'Balance' entails building up the exhibit so that it does not look top-heavy or lop-sided, and imbalance can be brought about by either getting the outline shape wrong or by wrong placement of blooms. The gerberas, for example, that could serve to provide a nice highlighting effect could, if bunched together at the apex of a triangular arrangement, make it seem that the top is about to break off from the exhibit. Balance is, in fact, created as work proceeds in building up the exhibit, and a critical eye must continuously appraise progress. It does not mean that identical material must unfailingly be placed at adjacent positions on either side of the vertical or horizontal axis of a design whether that be a Hogarth curve (a snaky 'S'-shape) or a Crescent or whatever else (Figure 7.1): but the visual state of stability which can be achieved by balancing colour, texture and form of blooms demands practice in the art, and as a beginner it would be advisable, I think, to arrange exhibits that are not over-ambitious in this respect. Asymmetrical arrangements should perhaps therefore be tackled 'later on' when some experience in producing symmetrically balanced exhibits has been gained.

Harmony

'Harmony' supplements 'Unity', and it entails arranging an exhibit that is pleasing to the eye. Concord between

Figure 7.1
Flower arrangement
employing 'Hogarth
curve'. (a) Down-
ward sweep of
blooms too
emphatic; (b) a
nicely balanced
arrangement

(a)   *(b)*

components in respect of form and colour must be felt by the exhibitor much as a painter senses that the scene on the canvas 'looks right'.

Thus in Floral Art there are some guiding principles concerning design that will help basically to 'get things right', but individual talent must hold sway. One helpful point concerning the flower arrangement section of a show is that usually flowers do not have to be grown by the exhibitor, and hopefully those who are not able to grow their own will have friends and neighbours wishing to provide at least some of what they need; because flowers can be expensive to buy, particularly the ones that you feel you need to create the masterpieces of your choice!

Whatever the origin of the plant material, it must be treated with care, and having kept flowers and foliage fresh in containers of water, it must be conveyed to the show without damage; so everything that is packed in boxes must be protected in the same manner as for items in the cut flower classes. In fact, flower arrangers' ideas on preserving freshness are those that can best be followed by other exhibitors in most cases, and thus it would be expected that hard stems would be slit upwards for about 2.5-5 cm (1-2 in) at the bottom, or a similar amount of bark scraped away from woody stems. Hollow stems will be cut under water, and filled with water and plugged with cotton wool, and cut stems which exude latex will be briefly cauterised with a match or candle flame. There are other 'tricks of the trade', and one particularly useful practice is to wrap tulips tightly in moist paper so that the stems are straightened; and warm hands and mild pressure will work wonders in inducing required shapes from flowers and leaves that have not grown to the exhibitor's liking.

There is one matter of some contention, that I think I must mention: it is the question of whether exhibits have to be 'made on the premises' of the show or whether it is permissible to make them up at home. The latter practice is certainly frowned upon in many quarters, largely because there is no check that it is the competitor's own work, and if the show schedule makes it clear that arrangements must be seen to be constructed by the competitor concerned without a major contribution from someone else, then there the matter rests. There are, however, attractive items that are easy to make up in the comfort of one's home, thus permitting valuable time on show day to be devoted to other exhibits. The making of a lady's corsage is an example. This requires the deft use of fingers to bind several blooms and a little fern or other foliage with florist's wire and gutta-percha into a handsome-looking spray and could more easily be made up at home and kept fresh by having the ends of stems in a little water and with the leaves in contact with a slightly damp cloth. The corsage will have been taken to the show in a suitable container, such as a cotton wool-lined cellophane box; all that is then required is to finish off the binding of the bottom ends which should have been dried off. A fairly well-known flower that is used by florists is the chincherinchee, the separate blooms of which will stay fresh for more than a day without being in water. These smallish blooms can be wired by the tiny stalks and used in conjunction with small ivy leaves which are also wired – through the back – and the wire is covered by gutta percha. This requires a slick rolling method to twist the gutta percha quickly round the wire, but practice will soon facilitate this to be done without trouble. But the making up of sprays or corsages, at home or otherwise, must comply with the local show schedule.

The potential numbers of exhibitors for the Floral Art section of flower shows can be very high indeed, because the skills involved are being practised every day, so to speak. The differences between restricting one's expertise to making up arrangements for the home and exhibiting them at shows lie in the slight extra effort that is required to get things to the show venue and in complying with the rules of a schedule. I know of few people who have not found the effort worthwhile, not so much because they may have won prizes – although this helps to boost one's morale – but because of the real enjoyment of the occasion.

8. The Domestic, Handicraft and Wine Sections

The Domestic, Handicraft and Wine sections of a show are usually popular, and the range of classes provides just about something for everyone. The Domestic section covers foodstuffs – bread, cakes, jams, pickles, etc. Handicrafts include embroidery, marquetry, painting and sketching, photography, and many other subjects pursued as hobbies or spare-time activities. Wine usually has a section to itself, and home-made beer is included. For all the items exhibited in these sections, however, the basic guidelines are the same as those applicable to the Horticultural sections: that is, everything has to be staged attractively and must therefore have a clean 'finish' and be neatly displayed on the show bench. And, of course, the schedule must be complied with.

The Domestic Section

I have to assume that would-be exhibitors in these classes for home skills will be basically equipped to do so, both with knowledge and tools of the trade, but perhaps it would be useful to run through some points that need to be watched. Taking first the Domestic section, some thought must be given to packing items for conveyance to the show. Whilst it is fairly simple to prevent breakage of jars of jams and pickles by packing newspaper around them in their boxes or baskets, some other items can be spoiled by just a little jolt in the car or by not keeping their containers upright. I am thinking particularly of cream gateaux: for them, movement in the tin can spell disaster. A usually employed method of 'packing' is to place the gateau on the lid of a deep tin so that the body of the tin can be carefully lowered over the top, to fit inside the lid. This leaves, or should leave, a space of about 2.5 cm (1 in) all round inside to permit manoeuvring the top part of the container away from the lid at the bottom with no damage to the sides of the gateau; but this space is sufficient to allow sudden movement should the container not be securely wedged in a box or should it be tilted. So it is well worth taking precautionary measures to ensure that all the hard work of producing a delightful exhibit is not thrown away. The judge will not be able to award points for 'hard luck' !

Exhibits shown in glass jars must not only conform to the scheduled requirements for size – and hopefully these will be stated as 'approximate' – but also the jars must be crystal-clear and labelled with a neat description of the contents, including the date when bottled or made. The maximum mark that a judge can award for the 'container' is relatively low, but in close competition, a label that is askew could be the arbiter. As a matter of interest, a typical marking system for jams and other preserves provides for a maximum of 20 points which are distributed as follows:

	Points
Container	1
Colour	5
Quality and consistency	6
Flavour and aroma	8

Other items such as cakes and pastries are judged on a similar basis, although the table of points differs in its construction, as follows:

	Points
External appearance, shape, colour and uniformity	4
Internal condition, depth of crust and texture	4
Distribution of ingredients and flavour	12

I must warn, however, that you must not expect to find a card against every exhibit showing the points awarded, as much depends upon the rules under which a show is to be judged, and perhaps to some extent the local practice adopted by the judges concerned. Furthermore, the pointing system may be modified from time to time. What is set out above, however, will give a general idea of the qualities that should be present when the exhibit is placed on the show bench.

One point of note is that 'Uniformity' has a part to play where, say, a plate of six tarts or scones is exhibited, just as it counts when showing fruit and vegetables. It is therefore wise to exhibit a 'matching set' in respect of colour, shape and size. Also, the placement of the individual units of a composite entry should be done with care, the six tarts, for example, being set up in a neat group. Needless to say, too, a white doily on a plain plate will help to set off the exhibit; and a similar type of presentation should be adopted for fruit cakes, sponge sandwiches and suchlike.

As a further reminder, do check with the schedule what is

actually required. If a dimension is specified – 'baked in a six-inch tin', for example – it must not be thought that the judges will be lenient in regard to your seven-inch entry. Or, if the schedule contains a recipe to be followed implicitly and it is not merely shown as a general guide, then that is the recipe to be used. This is not really an onerous discipline to follow, and in doing so one can so easily insure against being judged 'Not as Schedule' which would be terribly disappointing.

The Wine Section

I will take Wine next because, like the cakes and jams, not to mention the pickles, it has a factor that is not common to exhibits in the horticultural classes: it is tasted as part of being judged. I have a note of a marking system used in judging wine, and two-thirds of the marks are awarded for 'Flavour, Balance and Quality'. The remaining one-third are divided between 'Bouquet', 'Clarity and Colour', and 'Presentation', the maximum marks for which represent only about seven per cent of the total; but, of course, should not be thrown away by lack of attention to detail, because, in a tight finish, they could be the deciding factor in winning first prize. Very often, the rules about staging wine are quite specific in regard to the type of bottle – clear and unpunted (flat-bottomed) possibly – and its stopper, which is usually required to be flanged, thus saving the judges' having to use a

(a)   (b)

*Figure 8.1
Neat labelling of
bottles (a) is
important*

corkscrew. Never mind that vintage red wine is properly kept in a 'dark' bottle; if the schedule asks for 'clear', which, of course, assists examination of the clarity and colour of the wine, a bottle of clear glass must be used. A nice label, and not a mere piece of sticky paper, should be neatly affixed to the bottle, showing the necessary detail (Figure 8.1).

The Handicraft Section

Long have I admired the skill that is evident when one looks at the Handicraft classes, but, of course, presentation is important if the exhibits are to get their rightful awards. On home-made garments no loose ends of wool or cotton must be seen, bearing in mind that when several exhibits of apparently equal merit are facing a judge, he or she will search out any such minor defects in order to arrive at a decision about prize-winners. By the same token, if art work is framed, the frame should be neatly made and chosen to enhance the picture or sketch; and photographs should be mounted with great care. It is possible that a painting, photograph or whatever else in the art line could be of such outstanding merit that a poor frame or rough and ready mounting would not prevent its getting the first prize, but the judges will invariably take exhibits as complete packages, and they are bound to regard any bad 'finishing' as a sign that the exhibitor has not done his or her best to win approval; so why should they not, when competition is keen, let the scales be tipped in favour of the competitor who has taken obvious care in 'presentation'?

In staging exhibits in the Handicraft section, a lot will depend upon the type of show bench available. 'Woollens' and other garments can be neatly laid onto an uncreased sheet of tissue paper, or, if possible, they can be arranged on a small stand. Pictures can obviously be better seen if they are propped up against a box, or better still, the sponsors of the show might provide a special backing to the show bench – perhaps peg-board. It is advisable to find out about this before the show, and take along whatever is thought to be needed and which will not meet with disapproval from the committee. Usually they will be as keen as you will be to see exhibits nicely staged, and you will, I am sure, get the utmost co-operation.

I think that great care will be exercised as a matter of course in packing everything for carrying to the show because hand-made articles are treasured and sheets of glass in picture frames always seem to be well padded with thick newspaper. But seldom do such a miscellany of articles get carried around together, as they do on show days, and it is wise to envisage the worst possible set of circumstances occurring, and to take adequate counter-measures. For example, a picture might be laid down, only for a second, on a table in the staging room or hall, whilst you return to the car or go to the committee table to collect show cards; and just further along the table, a fellow exhibitor might have an accident with a vase of water; your picture might be well enough packed to prevent glass breakage but not proofed against a soaking. Far-fetched this may sound, but it is surprising what might happen even if the chances are slight; and in the case of paintings and similar items, there is always the risk that rain will fall just as one arrives at the show venue, and it will then be regretted that a polythene wrapping was not placed round them at home, a job that would have taken but a few minutes. It really does take relatively little time to take a few simple measures to protect one's exhibits and so ensure that they can be displayed in a perfect state; and unless that is done the risk of damage is always there, and it is a risk that is not worth taking.

9. And Now What Have the Judges Done?

Having successfully conveyed all one's exhibits safely to the show venue and staged them to the very best of one's ability, there comes the nail-biting period when all exhibitors are excluded from the hall or marquee and during which the judges, accompanied by stewards, will go about their task. In practice, a lot of exhibitors at local shows may have far too much to do locally, and thus there may not be time to worry very much about the outcome of one's efforts. But of course there will be the urge to see at some stage what success has been gained, and whilst for many the domestic necessities may take over for a while, many others can hardly wait to get back into the hall to see what success they have had. As the average period between the completion of staging and the show being opened to the public is something of the order of two to three hours, the alternative to going home or fulfilling some other local commitment is to relax with a pot of tea and some sandwiches or to have a more sustaining lunch if that has more appeal. At some of the major shows held in the RHS halls, the period of waiting is sometimes only about an hour and a half, because teams of judges are appointed in sufficient numbers to perform the judging within that relatively short time-frame, a 'luxury' that is beyond local resources as a rule. Whatever the number of hours, the period of waiting has to be accepted; and clearly, however keen one is to see what judgements have been made, there is nothing to be gained by worrying, and wherever possible, the time should be spent 'relaxing'.

Unfortunately, when you are at last able to go back into the 'arena', some of your expectations may fall short of the mark, and to your immediate disappointment, the awards which the judges have made may have gone to exhibits of rival competitors. It is not at all unreasonable to wonder if the judges have erred, but the important thing is not to be upset, and instead to look critically at the other exhibits in comparison with your own. Exhibits must not be handled, so it is not possible to search for some of the faults that the judges may have discovered: but none the less, a logical appraisal can be carried out. Where exhibits comprise more than single numbers, 'uniformity' is worth a study, and it might have to be admitted to yourself that another competitor's dish of pears contains specimens that make a perfect match, although the quality might appear to be no more than equal to your own exhibit. If you can be fair, it is not difficult to accept that the best person has won, but there may have been some very close decisions and it will not always be possible to see precise reasons why certain exhibits have been judged as superior to others. By carrying out your own inquest, however, a lot can be learned, including perhaps something about 'presentation', and it would be a wasted opportunity not to do this wherever there is time. Usually there will be plenty of people willing to discuss points with you, both exhibitors, and also spectators, anxious to learn more about the technicalities involved. From their point of view, the more knowledge that is gained, the more enjoyable will be their visits to shows, although I confess that there are many who just like taking in the general scene and admiring what they see before them – particularly, perhaps, those juicy black grapes that simply ask to be eaten. No matter that the bunch is a little crooked, and may for that reason have been beaten into second place; the huge fruits with their lovely 'bloom' simply look lovely, and that is enough!

Hopefully then, in half-an hour or so after going back into the hall or marquee to note the awards that have been made, you should have been able to reconcile your doubts and thoughts and made a firm resolution about the additional measures to be taken next year to guarantee success. If, however, you have found what seems undoubtedly to be a mistake by the judges – for example, the placing of a competitor's exhibit ahead of yours when it appears that the former breaches something in the schedule – you can consider putting in an appeal. Usually there is a simple but strict procedure stated in the schedule for this, and very often an appeal has to be lodged with the committee within an hour or so of the completion of judging. It is to be hoped that appeals will be rare – and indeed they are. They can give little joy to hard-working committees at the end of a tiring day; but, there it is, a procedure has to be available for ensuring justice in cases where there are real grounds for complaint. However, I rather like the advice in the RHS Handbook to the effect that exhibitors should accept all decisions of the judges with good grace.

The final stage comes as the show is 'broken-down' at the end of the day, and everything has to be taken from the show benches and the benches themselves stowed away as well as all vases and other equipment. Very often, produce that is not required by the exhibitor will be auctioned, and flowers are taken to local nursing homes or to the church. Usually the exhibitor will indicate by a 'reserved' card that certain items are not to be disposed of, or at least he or she will need to be on hand when everything is being gathered up. Probably one has a light-hearted feeling at this stage, not least because it has all been felt to be worthwhile, and perhaps to some extent because packing up can be done with less effort than when everything had to be so carefully packed for the show. Even so, there will be many exhibitors who wish to take some prize specimens to another show – giant onions, for example – so they will have to be carefully re-packed into their box. And so, of course, will handicraft items. All in all, however, much has been gained from the work put in: the committee will have felt pleased that they have organised yet another grand event, and the exhibitors – and some spectators – will have renewed friendships or made new ones. Few, if any, should feel dissatisfied, and there should be a firm resolve to attend next year's event in one capacity or another.

10. A Selection of Flowering Plants for Exhibiting

In chapter 3, I ranged over the possibilities of producing horticultural exhibits for shows, bearing in mind that one's activities in the 'flower show world' ought to be compatible with everyday responsibilities. In this chapter, I have drawn up a list of those flowers which I think provide a beginner with sufficient scope for exhibiting at the various times of year when shows are likely to be held. (Further details on frost tolerance, etc. can be found in Appendix IV.) Some demand more time than others if a high standard is to be aimed at; but they are all grown to a degree of excellence in many quarters, and quite commonly by many who have yet to enjoy the pleasure of showing their quality blooms in competition.

Annuals

Annual flowers are strictly those that grow from seed to maturity, flower and set seed, and then die within approximately twelve months. This strict definition rules out those kinds of flowers which we grow each year from spring-sown seed, to provide a mass of colour in the summer, but which, if we did not discard them, would not necessarily die. Sometimes a show schedule will spare the exhibitor worrying over this point by making it clear that all the customary seed-raised bedding plants that are discarded in the autumn/winter after flourishing during the summer months are to be regarded as 'Annuals', but more often just a few of those will be specified, including the antirrhinum. A further difficulty is that seed catalogues may advocate growing a certain plant as an 'annual' whereas it may have a perennial character of being able to survive for more than two years. The RHS recommend, in fact, that instead of having a class for 'Annuals', there should be a class calling for 'Flowers raised from seed during the twelve months preceding the show'; this would include some kinds of plants that are grown as 'Biennials'. An alternative would be to revise for show purposes the definition at the beginning of this paragraph by including 'or are commonly discarded' as a qualification after 'die'. Until new standardised definitions are drawn up, however, we must accept that show schedules may vary in their outlook on the subject of 'Annuals', and it is important to check on this point in every case.

It is customary to raise half-hardy annuals 'under glass' in the early spring or, in some cases, later in the spring, thus covering a period from the beginning of March until the end of April, dependent upon the 'greenhouse' facilities that may be available. A guide to temperature requirements is readily available on seed packets, but the important thing is to ensure that the seedlings make steady progress, and this may very well mean waiting a while before sowing if the heating, say, in the greenhouse is geared more to just keeping a few degrees of frost at bay than to raising day and night temperatures to the stated minima of perhaps 13-16°C (55-60°F). Probably one of the most cost-effective methods is to use a soil-warming cable that will provide the 'bottom heat' that inspires germination and also adequately raises the air temperature in the propagator box in which the cable is installed. I personally found the purchase of such a device, which was supplied complete with a transformer to reduce the mains voltage to a 'safe' 12 volts, one of my 'best buys'. It heats a large home-made box which will house many smallish plastic tubs or fairly shallow (about 7.5 cm (3 in)) pots, each of which will provide a host of seedlings grown in a soilless compost. The obvious difficulty remains of ensuring a suitable temperature in the greenhouse from the time the seedlings are 'pricked-off' into boxes which then require much more room than the propagator can provide. A simple double-burner paraffin lamp under the staging helps a lot, but I am afraid that with only these elementary aids, it is necessary to curb one's impatience and not sow before the end of March – and later still in 'the cold North'.

Hardy annuals probably offer more scope for early summer shows because the seed seems to germinate easily at lesser temperatures than demanded by half-hardy kinds. But, although they are often grown from seed sown *in situ*, and for some kinds a September sowing is recommended in order that an early display can be had from the over-wintered plants, I think that sowing thinly in boxes, or, better still, sowing one or two seeds in pots, will be a more successful plan if some good blooms are required for a summer show. This can be done in a cold greenhouse in early March or April depending upon the local weather conditions.

When the time comes to plant out the annuals, the ground should already have been prepared by digging-in some peat and any available well-composted organic material, and a dressing of fish, blood and bone fertilizer lightly raked-in; about 60 g (2 oz) per sq yd will suffice. The time will be dependent upon the weather, with hardy annuals being planted from early April onwards and the half-hardy kinds having to wait until May, and very possibly the end of that month. For those with less space available, however, the size of bed need not be so large, and in a border of clay or gravel it might be possible to dig out an area of about 3 sq m (3 sq yd) to a depth of 15 cm (6 in) and fill it with a mixture of peat, loam and composted manure; this would enable, say, ten plants of several kinds of annuals to be grown with some success.

Very often, local shows have separate classes for single and double-flowered varieties (cultivars) of this half-hardy annual; and although it is not the easiest of flowers to grow successfully, suffering as it often does from wilting, the seed is quick to germinate if a modest amount of 'bottom heat' can be given, and the plants will develop fairly rapidly. A March sowing will be necessary if a July show is in mind, and pricking-off the seedlings should be possible within five or six weeks of sowing. They must then be grown on as fast as possible until being 'hardened-off' in a cold frame or by being placed in a sheltered area of the garden by day – perhaps on a patio – and protected at night. Planting-out will need to be done before the end of the third week of May, but in cold spots it will be necessary to wait until the beginning of June – in which case, hopefully there will be a convenient show in August or early September! It really does become a matter of experience to determine what is, or is not, possible to achieve with any degree of success. But asters are delightful flowers and well worth at least a 'try': of the double kinds, I prefer the Ostrich-Plume cultivars, seed of which can be obtained to produce a good range of colours. Long stems are desirable, and I think it helps in producing these if regular liquid feeds of Maxicrop or Phostrogen, or a similar fertilizer that can be applied as a foliar feed, are given at ten-day or fortnightly intervals, once the plants are established in their position in the garden. Growing them strongly in this way might also help to keep wilt disease away.

Asters

The stems should be cut to their maximum length, the ends slit, lower leaves stripped, and then stood in water; one or two days before the show should be acceptable to the blooms provided that they are just approaching their peak when cut and are kept in a cool place. Unless it is possible to carry them to the show wrapped in tissue paper or in a container of water, the stems can be safely packed into flower boxes, with their heads protected by tissue paper. To prevent movement of the stems, they can be held in place with pieces of split cane wedged across the box.

Staging the blooms should be kept on simple lines, and unless the schedule asks for something different – say, 'A vase arranged for all-round effect' – the stems should be set up in short rows, the front row of stems being of less length than those behind so that the blooms stand clearly separated from each other, but not with excessive gaps. If no number of stems is specified seven stems in rows of four and three, with the front blooms sitting in the spaces below and between the top ones, should do nicely. Sphagnum moss will enable the arrangement of stems to conform with the exhibitor's wishes, or paper can be used and a little moss placed in the top of the vase to conceal the unattractiveness of paper; or a water-retaining material such as oasis can be used provided that it fits snugly in the vase and not too many attempts are made to get things right. Otherwise, the material will tend to crumble.

Calendulas
(Pot Marigolds)

Unlike the African and French marigolds, the so-called pot marigolds, botanically named calendulas, are hardy annuals, and I thoroughly recommend them for a summer show. For garden display, in districts without harsh winters, seed can be sown in the open in September, for producing flowering plants in late May or early June; but for show purposes, I think it is better to sow two or three seeds in some 7.5 cm (3 in) pots at the beginning of March or slightly earlier and place them in a cold frame or greenhouse. From the beginning of April, given that it is not then snowing or freezing hard, the pots can be stood outside, and the plants can be bedded-out as soon as convenient, or they can be potted-on into 12.5 cm (5 in) pots to provide some nice pot plants. This should enable blooms or pots of flowers to be shown in July, but it is worth experimenting a little with different sowing dates and treatment thereafter particularly as temperatures fluctuate wildly in the springtime, with cold greenhouses getting very hot in March days and very cold on April nights. A steady temperature of around 13°C (55°F) would be desirable, but without temperature control equipment, one must simply do one's best; and it never ceases to surprise what can be achieved by some rudimentary measures and determination.

The stems for showing should be cut as long as possible, lower leaves stripped, and then placed in water. When picked at their prime, the blooms will last satisfactorily for one or two days in a cool place. There is a fairly wide range of cultivars available from many seedsmen, but I think that a tall cultivar such as 'Orange King' which has lovely double, glowing blooms, should be chosen for an eye-catching display. If packed into flower boxes, the blooms must be protected by tissue paper to avoid their becoming misshapen. Staging can be as for asters.

The hardy cornflower will resent mollycoddling even more than the calendula, and it would be no bad thing to try two methods in parallel: an outdoor sowing in September which could produce flowers for cutting in June, and some seeds in pots as for calendulas. Although there is a mixture of colours available, I think that the large-flowered blue cultivars will make the most impact on the show bench. It is best to delay cutting the stems until the eve of the show because the blooms seem to lose their freshness rather quickly. Staging can be as for asters, but as the blooms are relatively small, the spaces between them must be reduced in proportion; and, unless a specified number is requested, a dozen stems, arranged in descending rows of five, four and three, would be appropriate. *Cornflowers*

Known also as the annual delphinium, and hardy, larkspur will produce spikes of bloom that are likely to be the most spectacular among annuals at a show. It will benefit very much from being grown in good loamy soil, and 'thin' soil cannot really be expected to produce majestic spikes, although a liquid feed at ten-day intervals should be given in any case. As with most plants, acid or alkaline extremes are not favourable. A similar pattern of seed sowing to that suggested for cornflower should be followed, and hopefully the spikes will develop sufficiently to require staking, because as with delphiniums, a good length of bloom is required. *Larkspur*

The florets will start to wilt fairly quickly after the stems are cut, so cutting should not be earlier than the day before the show, and the usual long drink of water is absolutely essential. Careful packing, using tissue paper, is required, and in view of their length, about a dozen stems will occupy a flower box, but if six heads are placed at one end and six at the other, some of the shorter annuals could be fitted into the middle of the box. Seven stems, with three sitting comfortably in front of the back row of four, will make a nice vase; and the remaining blooms that are taken to the show will help to bring life to a vase of mixed flowers in a summer show.

Although larkspur is strictly a biennial, it is commonly accepted as an annual.

The cultivation programme for this half-hardy annual is the same as for annual asters, but it is more robust and seemingly hardier. Good light rather than intense sun is helpful as indeed it is for many annuals, although some of them enjoy *Marigolds, African*

71

having their heads in the sun more than others. Zinnias, for example, positively relish it. However, choice of site usually has to be a matter of compromise between a number of factors, and in practice based on the overriding requirement to avoid absolute shade.

Large flower heads are required, and there are several cultivars that will produce them, including the popular 'Crackerjack' mixture. Whether a single colour cultivar or a mixture is decided upon, tall-growing plants with long flower stems are needed, and certainly not a dwarf type which could very well produce large blooms – and there are many cultivars that will do this – but only short stems. When just fully open, the blooms will remain fresh for a day or so, provided that the usual procedure is followed of cutting and slitting the stems, and then, with the lower leaves stripped, placing them in deep containers of water. Packing into flower boxes should not damage the blooms if care is taken to protect them with tissue paper. Staging can suitably be as for asters.

African marigolds will bloom into the autumn, and two sowings in early March and mid-April could enable vases of these excellent flowers to be exhibited in both a summer and a September show.

Rudbeckia

Basically a perennial, this plant can be grown as an annual, that is to say, being discarded at the end of the season and, provided that the schedule does not exclude it, it can be a very useful plant for producing a winning exhibit of a vase of mixed annuals or of one kind. Seed of a mixed colour, tall-growing cultivar is a good choice, but a single colour cultivar by name of 'Marmalade' will provide a glowing exhibit for a summer show. If sown as though a half-hardy plant, the earliest possible start should be made, depending upon the available temperatures, assisted as needs be by gentle heat from a lamp or a soil cable. Early March is about the best that can be done in this regard unless more sophisticated environmental control equipment is used, and with any luck, blooms would be ready for a latish summer show. If treated as a hardy annual, plants for flowering in September could be raised from an outdoor sowing in April or early May.

Stems should be cut to maximum length as close to the show day as possible, and slit at the bottom before being immersed in water. Care must be taken to avoid creasing the petals which can be rather floppy, and the blooms should be cushioned in their box by pads of cotton wool or tissue paper. Staging should be executed on the lines already described for other annuals.

Sweet Peas

I feel compelled to mention in the short list of annuals for showing, the delightful sweet pea because it is so important to a summer show, bringing its own touch of magical beauty and fragrance: but to grow them well is none too easy. Strangely, I think that the method of growing them, cordon fashion, secured to 3 m (8-10 ft) canes is widely known and practised, and it is certainly no secret that tendrils are usually removed and the vines tied with soft twine or secured with special rings, to the cane. The aim is to produce lengthy stems of a thickness in keeping with the flowers and which carry four or five blooms evenly spaced (Figure 10.1). There is much general talk about 'fives', but the beginner should certainly be satisfied with stems carrying four blooms each rather than an assortment carrying differing numbers. There are, in fact, dwarf-growing cultivars such as 'Jet Set' which will provide long-stemmed multi-flowered specimens, but they do not seem to be in any position to challenge the supremacy of their taller colleagues.

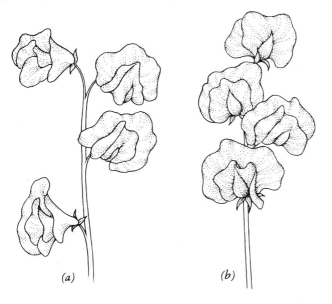

Figure 10.1
Blooms on sweet pea stems should be spaced evenly (b) and without gaps (a)

(a)   (b)

Raising the plants from seed is not difficult, and this can best be done by sowing them in pots in early October (Figure 10.2) – about four seeds in a 10 cm (4 in) pot containing a suitable 'compost' such as 'John Innes', or a soilless, multi-purpose type, of which several brands are available. A cold frame is very suitable, with protection being given as the weather becomes harder. Whilst the plants will put up with some frost, there is nothing to be gained from

Figure 10.2
Sowing sweet peas

subjecting them to unnecessary rigours, and the aim should simply be to let them grow in a hardy fashion; that is, not being mollycoddled. In a cold greenhouse, I have found that growth has been too fast, with the plants becoming too advanced before it is possible to plant them out. Sturdy, short-jointed plants are needed and not some carrying growth of 15-17.5 cm (6-7 in) in length; and experience will show how this can best be done. In fact, provided that the seed has been stored in a cool place and is 'fresh', the germination stage is probably the easiest part of the whole process, although some assistance can usefully be given to hard-coated cultivars, the seeds of which look very dark brown or black. These can be 'chipped' by removing a tiny sliver of skin with a sharp knife on the side opposite to the small 'eye', which will speed up the process of germination.

It is all well worth a trial, and even if top-class blooms are not obtained for a while, what joy there is to be derived from seeing a large crystal bowl of beautiful flowers on the hall table – for all callers to admire, – and from being able to keep it filled with fresh blooms which will perfume the house. If there is no burning desire to get into the top league in next to no time, it could be sensible to concentrate first on growing strong plants from an autumn sowing, and having mastered that, take the next step of cordon-growing – which is time-consuming. In other words, the plants could be grown up twiggy sticks or cylinders of wire held upright by canes. This would certainly produce lots of blooms, and some of reasonable quality, but not absolute winners and, of course, as the season progresses and the supports become covered by a mass of foliage, short stems will predominate. It is a matter of pacing oneself to suit one's own abilities and convenience.

In any event, the plants will need to be set out from their pots in March if possible or later if the weather so dictates. Initial protection against biting winds can be given by screening the plants with plastic sheeting – but allowing plenty of air to circulate.

Preparation of the site today seldom involves the former practice of digging really deep trenches and filling them with composted manure and so forth, but if one is physically able, it is wise to dig out to a spade's depth, fork the subsoil and put a substantial measure of composted organic material into the trench before replacing the topsoil – which preferably should be good loam. Alternatively you could settle for forking into the soil an ample quantity of peat mixed with a proprietary animal manure product of which there are many on the market. This preparation should be done well in advance of planting time to allow the soil to settle, and it can then be raked to produce a good tilth; and a week or so before planting, a dressing of Growmore or fish, blood and bone at about 60 g per sq m (2 oz per sq yd) should be raked-in. This is but an elementary outline of a fairly basic requirement for giving the plants a good start, and additional measures, including the application of bone meal or bone flour when the trench is dug can be found in a specialist book, as well as much other information on cultivation for exhibition and on cordon growing.

The supports, or at least some of them, will need to be in position before the plants are set out, and made firm against wind or storm by whatever method is convenient. The plants should be set out so that the bottom side-shoot is at soil level, and about 5 cm (2 in) to one side of a cane if that is the support being employed. Plants should be about 20 cm (8 in) apart in the rows and made firm, and twiggy pieces of stick will help to keep them upright; and slug pellets will be an unfortunate necessity.

When growing on a single shoot system, that is as 'cordons', it will be necessary to reduce the shoots to one by snapping off the tops of those not selected. Picking the best shoot will be a matter of experience, but in general it must look strong and vigorous, and often this is the case with the tallest shoot. Making this selection is better done several weeks after planting than earlier, and it follows that a raffia or fillis (soft string) tie around the plant and its support, enclosing any twiggy sticks, will be required to prevent its straggling onto the ground. As the plants progress, and tendrils are removed, tying at each leaf joint becomes essential, and later when buds appear, weekly applications of a liquid feed such as Maxicrop or Phostrogen should help, although some growers would hesitate to do this in the case of the stronger growing plants. Conversely some practise

Figure 10.3
The various parts of
a sweet pea stem

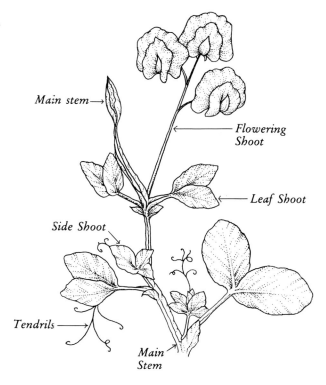

Main stem→

Flowering Shoot

Leaf Shoot

Side Shoot

Tendrils →

Main Stem

regular foliar feeds, using one or other of the fertilisers I have mentioned, from the time when plants are no more than 45 cm (18 in) high. Slightly more contentious is the question of disbudding, some exhibitors always removing the early buds showing perhaps when the plants are less than 60 cm (2 ft) tall, and others taking them off until there are signs of four-bloom stems developing; but this is not advocated by everyone, and there is even the argument that some cultivars known to be weak growers benefit more than others from disbudding until the plants are at least 90 cm (3 ft) high. To begin with, I would disbud up to a height of about 90 cm (3 ft) and always stop doing so at least three weeks before a show to ensure that there are some stems in bloom at that time! Later on in the year, plants will reach the top of their canes and a specialist book will explain 'layering', which entails lowering the plants to ground level and training them to climb canes further along the row, so that the horizontal length is added to the total height. Or, if showing is over and blooms are required solely for the house, the tops of plants can be pinched out!

Cutting on the eve of the show rather than earlier is to be preferred, and the ideal stage of development at that time would be a stem with the top bloom just about fully open, and the bottom bloom still nice and fresh. The stems must be placed in water immediately and kept cool. A few 'twin' leaves should also be kept fresh in water for use later as a nice 'finish' to exhibits. Various ways of carrying the flowers to the show are possible, the simplest being to wrap a bunch of a dozen or so stems in soft tissue paper and take them by hand – given fine weather – or the bunches so wrapped could be stood in empty vases or similar containers, several of which could be packed with crumpled newspaper into a box. If flower boxes are used, strips of greaseproof paper can be laid across the stems and a piece of tissue paper used on top of that to cushion the delicate blooms: this will avoid the possibility of the tissue paper absorbing any residual moisture from the stems, which should in any event have been wiped. With practice, it will be found that stems can be packed from one end of the box with the heads of blooms nearly touching each other, and then when the last row of stem ends reaches the end of the box, the procedure can be reversed in direction with the stems being tucked under the greaseproof paper.

Reeds packed tightly into a vase are used for staging sweet

Figure 10.4
Sweet pea vase
ready for staging
the stems

Figure 10.5
Vase of sweet peas,
stems neatly fanned
out

peas, and if available, they make the job of placing the stems into desired positions that much easier. A walk along a canal bank or stream may enable you to secure a supply, but failing this or any other source of supply, the familiar oasis can be brought into use. A spare piece of stem can be employed to make holes, in preference to risking breaking any of the stems needed for display. It is customary to insert at the back of the vase a twin-leaf, anchored by a short stub of shoot (Figure 10.4) and to then fan the stems out so the blooms do not quite touch. About six stems will look attractive if fanned out in one row (Figure 10.5), and two rows will be needed for eight or more, the front row having the stems shortened by 2.5 cm (1 in) or so to create a gentle slope towards the front of the bench. Twelve stems can be staged nicely in three rows of five, four and three, and stems in a bowl are customarily arranged to produce an all-round symmetrical exhibit with a slightly mounded shape from the centre towards the outside of the circle, the blooms being slightly separated and not bunched-up. A finishing touch to a vase is to insert a small twin-leaf so that it leans gently over the front.

Biennials

Biennials are those plants which require the seed to be sown in the summer of one year for the plants to bloom the following year. Two popular plants strictly classifiable as perennials, but which are commonly accepted for show purposes as biennials, are wallflowers and sweet williams. The former, in fact, is likely to be the only cut flower, other than those grown from bulbs, for a spring show.

Wallflowers

Raised from seed sown in late May or in June in the open ground and kept well-watered, plants are usually spaced out in a bed 15 cm (6 in) apart to produce stocky plants for transfer to their flowering quarters in October. A slightly alkaline soil will be welcomed rather than an acid one, and the presence of lime might help to keep club-root at bay, although I would advocate using a proprietary remedy when planting out. Powders are available in sachets for making up into a solution in which the roots are dipped – a simple process taking but a few moments. Planted along a wall, they will benefit from the radiated warmth and bloom ahead of those catching the cold winds that have an unfortunate habit of blowing in the weeks leading up to a show. An easy flower to grow, they are not given to presenting themselves nicely and will become quite crumpled if packed without protective tissue paper. If cut two days before a show when the bottom blooms on a stem are only just opened, and after the usual treatment of slitting stems and stripping bottom leaves, and placing the stems in water, the buds should open; but I suggest some experimenting to determine the optimum time. Certainly, if the flowers are well open when a chilling wind comes along, the spikes will not look their best on show day.

Nine stems in two rows of five and four will look adequate, and so far as colour is concerned, whilst there are some good mixtures available, I think that a single glowing shade of red as produced by the cultivar 'Fire King' is to be preferred.

Sweet williams will accept most situations in the garden except heavy shade, but they seem to be a favourite for growing on open allotment sites. The cut blooms are noted for lasting up to two weeks indoors, but I would not recommend cutting them farther in advance of a show than about two days, and when chosen it must be seen that none of the florets has started to shrivel. A system of cultivation similar to that for wallflowers should be followed, leaving out the club-root treatment, and a giant Auricula-eyed cultivar should, in my view, be chosen in order to provide a fetching exhibit. It is important to choose heads of bloom that match in size and shape – which should be circular – and which have no gaps between florets; and the same treatment for ensuring that they receive a good drink should be followed as for wallflowers.

Sweet Williams

If no specific number is asked for, seven stems will do nicely if arranged in two rows of four and three on the lines described for annuals such as asters. I must admit that I often encounter sweet williams staged in an all-round arrangement with their heads touching, and there is something homely about this; but I think that there should be space – not too much – between the blooms, and I prefer them facing the front of the bench if that does not contravene schedule requirements.

Perennials

The term 'Perennials' is used rather loosely in some show schedules when what are required are flowers from 'Hardy Herbaceous Perennials' which lose their above-ground growth in the winter to survive for more than two years, with new growth appearing each spring. In fact, some retain their basal foliage and do not completely die down, but the important factor is that none of them forms a persistent woody stem. Any plant that lives beyond two years, however, can be said to be 'perennial', and a whole range of genera is involved, including trees, shrubs, herbaceous plants and many others. If there is any ambiguity about the wording of a class incorporating the term 'Perennial', a word with the show secretary will be needed.

The following short list of plants that are grown as perennials should give a good showing and provide some attractive exhibits in summer and autumn shows.

Achillea

Seed of this herbaceous plant is available, but I recommend buying ready-grown plants, and preferably of a large-flowered cultivar such as 'Gold Plate'. Two years after planting, some nice long stems should be available for showing, but as regards an optimum time for planting, I think that as with most herbaceous plants, the choice between spring and autumn will have to be made as an individual judgement. In general, however, cold soggy ground is not suitable, and where this is the customary situation at the end of winter and running well into the spring, autumn planting in warmish, moist soil, with plenty of peat to nestle into, would be the choice.

The achillea prefers a sunny position, and flowers should be ready for cutting in July or August; but they last well and can be fresh in Septmber in some cases. If an early start in gathering exhibits for a show is particularly necessary, the achillea blooms should oblige by staying in fresh condition for about three days, after the stems have been slit at the bottom and placed in deep containers of water. Packing into flower boxes should not damage them if protection is given with tissue paper. Seven long stems with heads of uniform size and round shape, arranged in rows of four and three should look quite spectacular on the show bench, particularly if the heads are of 15 cm (6 in) diameter which is attainable when the plants are well-grown.

Delphiniums

As among annuals, I felt obliged to include sweet peas because of their special qualities, so I think that the delphinium must be listed as at least a possibility in a short selection of perennials because it has a unique majesty of its own. Normally hardy, except, of course, against possible ravage from slugs, it will not provide the exceptionally fine spikes that are to be seen at summer shows without some special attention in cultivation. This, none the less, is not at all over-demanding or time-consuming. There is a choice of buying plants of cultivars known to do well in the show world or of sowing seed of a selected mixture with the chance that one or two of the plants so raised will turn out to produce winning spikes. Perhaps, if delphiniums are not already being grown, it would be wise to buy in a few plants as 'bankers' and try seed-raising as well. Both plants and seed should be purchased from a specialist grower and seed of very good quality can also be obtained from the Delphinium

Society. This can be sown early in the spring in pans or seed boxes, and it is very important to keep the compost moist without fail. A modicum of bottom heat will start the germination process in March, but excessive heat must be avoided, and it will probably be found that daytime temperatures in a cold greenhouse, for example, will soar uncomfortably high in springtime, thus calling for plenty of ventilation. Seedlings can be potted singly in 7.5 cm (3 in) pots when two true leaves have appeared, and the plants can be set out in the garden when they look sufficiently robust. This could be during the summer months when the soil becomes dry rapidly, and if the plants are filling their pots with roots and are in need of moving on rather than being kept back for a cooler spell of weather, lots of watering will be required. A sunny site is highly desirable.

Delphiniums flourish in soil that has a good humus base, formed by digging in plenty of material from the compost heap, and they will drink copiously in the summer; but slugs will take advantage of the wet environment and seek out the young shoots, particularly as they begin to form in the early part of the year. Some form of slug bait or killer is, therefore, a necessary aid to cultivation if disappointment is to be avoided; an old-time practice was to scrape away soil from the base of a plant and fill the depression with weathered ashes on which the slugs could not 'slime', but these are not likely to be available to the majority of people today.

Planted in soil well-endowed with composted organic material as suggested, the plants should develop strongly, but a feed of fish, blood and bone fertilizer or an alternative general-purpose one, in the spring will assist growth, and liquid feeds at fortnightly intervals as spikes of flowers appear can also be given. The spikes will require staking to prevent the wind playing havoc with them, but equally the ties must not be made fast in such a way that a stem is held rigid at a point immediately below the flower spike with the result that it can snap at that point; it may therefore be preferred to have three or four canes round a plant and embrace the stems with one or two loose ties of soft fillis.

As with many perennials grown as border flowers, it is worthwhile raising new plants, say, every three years or thereabouts so that the old stock with its rather wizened crown can be systematically discarded. This can be achieved quite simply by severing new shoots in spring as they grow from the crown. They should be about 7.5 cm (3 in) in length, and, if it is not wished to lift the plants out of the ground, soil can be scraped away to reveal the point at which a shoot joins the crown; with a sharp knife, a clean cut will neatly secure the shoot for potting (Figure 10.6). Shoots will form roots readily in a mixture of equal parts peat and sand,

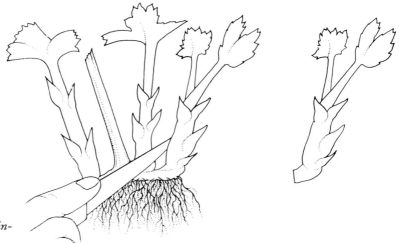

Figure 10.6
Taking a delphin-
ium cutting

or in vermiculite or a potting compost, provided that they are not allowed to dry out. This can be achieved by enclosing them in a little plastic world of their own, whether it be a large plastic bag (with some pinholes) placed over a flowerpot or a propagator unit consisting of a tray covered by a tallish, ventilated, plastic dome. The reader will probably be aware of a particular system that ensures success, the aim being to reduce moisture loss without creating excessive dampness which would cause rotting ('damping off'). The alternative of repeatedly watering the compost is a poor one because it will probably cause the bottom of the cutting to rot; and indeed, it is hoped usually to be able to avoid watering after the initial soaking of the potting medium for a lengthy period, and, if possible, until the cuttings are rooted. Of course, there are more sophisticated ways of striking cuttings, 'mist propagation' being one in which a fine mist of water is automatically sprayed over the cuttings, to reduce transpiration, and this also entails control of temperature; but simple measures can be highly successful with a little practice, and one soon becomes expert at judging the condition of cuttings and whether they need assistance – perhaps from a little 'bottom heat' if that can be made available – or whether they look as if they are 'happy'.

If cutting can be done on the day before the show and the flowers then given a good long drink, having had their stems defoliated, filled with water and then plugged with cotton wool, they should look fresh at the show provided that the bottom florets were showing no sign of flagging at the moment of cutting. A deep container weighted down with some large stones will be needed, and it must be seen that the stems do not topple over the side onto the floor. Several alternatives to ensure this are possible, ranging from holding them in position with some of the stones, taking care not to bruise the stems, to steering them down through holes in a fitted lid. The odds are against being able to convey them in their containers to the show because of the height that would be required in the vehicle concerned, let alone the space all round. Specially constructed boxes are used by some experienced exhibitors, and others carry the spikes wrapped in florists' paper, but florists' cardboard boxes can be used to hold a few blooms each. Holes will be required in one end to accommodate the bottom inches of the stem which will thus protrude, and tapes secured through slits in the sides of the box and held taut about 7.5 cm (3 in) from the bottom (Figure 10.7) will enable the stems to lie so that the flowers are not pressed on the unyielding cardboard; they can be given additional protection with tissue paper. Alternatively, the spikes can be carefully wrapped in florists' paper before being placed in boxes. If a box is to be carried by hand to the show, it will be necessary to wedge the stems inside with short pieces of split cane wrapped in tissue paper, but in the case of conveyance by car it should be possible to keep the box on an even keel more easily, and the paper protecting the blooms may be adequate. This is very much something that practice will perfect.

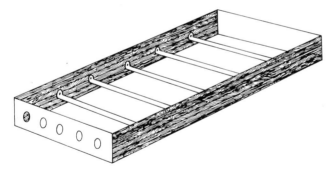

Figure 10.7
Flower box adapted
for carrying long
flower spikes

In staging the blooms, the base of each spike should sit above the top of the vase by between 10-15 cm (4-6 in) so the stems will need to be shortened accordingly, and refilled with water and plugged. The laterals should have been removed when cutting, but if any small ones remain through oversight, they should be taken off before the stems are arranged in the vase. As a matter of interest, however, I ought to point out that the RHS Handbook states that no differentiation should be made between spikes shown with or without laterals.

A single spike should be bolt upright in its vase, and this

can be ensured by packing it with crumpled paper topped with some sphagnum moss, into which one or two leaves can be placed to provide a nice finishing touch. In the case of exhibiting three spikes in a vase, a slight lean outwards of the two flanking spikes will be necessary, but this must not be exaggerated (Figure 10.8, Plate 11).

Figure 10.8
Vase of three flower spikes staged (a) too close; (b) correctly; (c) with exaggerated tilt

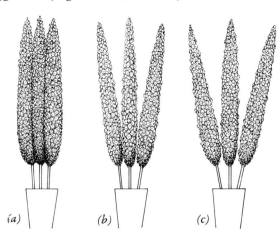

(a)   (b)   (c)

Gaillardias

Gaillardias give a handsome return for a modicum of attention. They like a well-drained site in a sunny position and can be raised from seed sown in June, the plantlets being pricked-off into a 'nursery bed' as for biennials and then grown on until ready for planting in their more permanent quarters. In fact, perennial as they are, I think it is wise to sow every three years or so to safeguard against inevitable deterioration in the older stock. Just half-a-dozen or so plants will yield as many stems as are likely to be required for showing, and the period of flowering extends from summer into early autumn, making the gaillardia a useful plant for both a July and a September show.

Stems can be cut with the flowers fully open and kept for a day or so in water in a cool place. Seven stems of the large-flowered hybrid cultivars will make an attractive exhibit.

Michaelmas Daisy

There has, I think, been some loss in popularity of this hardy plant because over recent years mildew seems to have become widespread. Although the group concerned, named *Aster novi-belgii*, contains some really lovely cultivars – 'Fellowship', for example, with its very large blooms, fully double – fortunately there are some nice cultivars in the *novi-angliae* group which do not suffer from mildew. 'Harrington's Pink' is a well-known cultivar in this group, and its shrimp-pink flowers are a joy to behold: and some modern cultivars are now producing quite large blooms. There is, too, the *Aster amellus* group containing the famous 'King George' cultivar of a lovely blue with yellow centre. 'Guaranteed' free from mildew troubles, this cultivar has so often been a winner in the past.

Michaelmas daisies tolerate various garden conditions, and dappled shade will suit them. They do not last fresh for long after being cut despite having their stems slit, and I would counsel cutting the stems no earlier than the day before the show; and stems carrying tired blooms should not be selected, although a few blooms could, if no alternative stems are available, be carefully snipped off with scissors. It is not an offence judiciously to remove blooms or florets from stems of flowers exhibited – unless there is a specific rule against it – but where it is noticeable, it must be expected that the exhibit will lose points. So it will if 'spent' blooms are left for the judges to witness! Entirely fresh flowers must therefore be the prime aim.

Michaelmas daisies need protection from being crushed if packed in boxes because little or nothing can be done to uncrumple creased blooms. Seven stems in rows of four and three would need a large vase, and just five stems, well-clothed with flowers, ought to suffice. But, check the schedule, as always, for numbers required; it might well be that only three stems are requested.

Pansies

Although these are often grown as biennials or half-hardy annuals, and are recommended by nurserymen on this basis, it goes against the grain to discard them all so quickly, and if spent blooms are removed unfailingly and the plants tidied up at the end of the season, it is possible to get some reasonable blooms for two summers, after which they can be replaced by new plants grown from seed. The plants are fairly accommodating in their requests, but they dislike intense shade, and will appreciate being tucked-up in some peaty soil and regular liquid feeds of Phostrogen or Maxicrop.

The blooms will need pre-conditioning by having their stems immersed in water for the whole of an evening before a show, because the customary method of staging is to place the stems into a container of damp sand, or, as is now becoming popular, a low bowl holding a piece of oasis. Without the long, deep drink beforehand, the blooms will show signs of wilting, particularly in a hot environment. It is important that the blooms should not overlap in their container, and staging, say, ten blooms is best done in my view by setting them across the container in rows of three, four and three. It should, at least, be of some relief to enter a class for pansies because there is none of the usual problems of weight to be carried or length of stem to be coped with, but this does not entitle lack of care, and the blooms should be gently handled by their stems and laid on cotton wool in a trug or shoe box. A container of stems of a cultivar producing mixed colours with marked 'faces' provides, in my view, as nice an exhibit of pansies as an exhibit of self colours, but what is admissible will depend upon the wording of the schedule. Fortunately, there is usually the widest scope, because most local societies ask simply for 'a bowl (or container) of Pansies' with a certain number of stems stated; but should the terms 'Show' or 'Fancy' Pansies be encountered, I would advise reference to the RHS Handbook.

Scabious

There is a lovely powder-blue cultivar of the Caucasian scabious named 'Clive Greaves', and a vase of nine stems of this beautifully delicate flower will always be a strong competitor. Plants purchased in the spring should produce some good blooms in the summer of the following year. A lightish, slightly limy soil is preferred, and a sunny border is the best situation. New, vigorous plants can be established from small 'divisions', that is, pieces of younger plant eased gently away from the outside of the crown by use of a handfork. These divisions should have plenty of fibrous root, and in the spring they should establish themselves quickly to produce within 15 months strong plants that can take over from some older ones that may be going into decline.

I think that the delicate heads must be protected with cotton wool when they are packed for the show, and cutting the day before the show is preferable to doing so earlier because absolute freshness is the keynote.

Roses

I think roses fit into a section of their own although they are perennials and also shrubs. Not only is there such a range of them, but they are also a 'Queen' among flowers and there are Rose Shows all over the world. Old names persist, albeit such terms as 'Pillar Rose' seem not to be known with the affection of yester-years: but a reclassification is taking place to produce a nomenclature that can apply internationally. Members of the Royal National Rose Society will be aware of these changes in description, and I think it could be said that the two changes that will be of major interest to local exhibitors are the introduction of the terms 'Large-flowered', embracing the previous 'Hybrid Tea' terminology, and 'Cluster-flowered' in place of 'Floribunda'. The RNRS have themselves simplified matters by stating in their rules for exhibitors that the criterion for classifying roses should be whether the exhibit meets the standards that they have clearly set out. Thus, for example, a bloom cut from a modern climbing rose and of appearance that matches the description concerned – which requires it to resemble a bloom commonly exhibited under the 'HT' section of a schedule – would be eligible as a 'Large-flowered' exhibit. Similarly, a nice cluster of blooms on a stem cut from a 'climber' could be exhibited alongside a stem of blooms from a floribunda rose, both of them qualifying as 'Cluster-flowered'. Until the new terminology supersedes any existing wording in any particular schedule, however, the latter must be adhered to by the exhibitor.

The best type of soil for roses is 'a rich loam': which over the years seems to have become translated in error to 'clay'. How often, even today, does one hear someone say 'roses like clay' when what is really meant is that they do well in a soil that is akin to a medium to heavy loam rather than a thin soil that has little body. Relatively few people, however, are fortunate enough to have 'textbook-perfect' soil, and digging in plenty of 'compost', peat, rotted manure and so forth will usually be necessary to ensure a good fertile bed for planting roses. Now that we have a proliferation of Gardening Centres, the term 'container-grown' has become fashionable, and the popular theory is that most plants can be planted from them at various times of the year; and to a large extent this is correct because there is relatively little root disturbance when a plant is knocked-out from a container and planted with a good ball of soil, as an entity, and it should not take long for the plant to settle in. But it must not be ignored that roses must not be allowed to become dry at the roots if they are to succeed, and if, as a matter of convenience, they are purchased for planting during a dry and sunny spell, they must be well supplied with water at planting time and thereafter until they have established themselves in their new environment.

Having said this, I must confess that I am perfectly happy with 'bare-rooted' plants dug from the fields of the nurseryman, which I plant in November in compost enriched soil, first digging a hole wide enough to take the width of the roots which are placed into several inches of the soil mixture and then covered so that the union of briar stock and the budded growth is just about at the level of the top of the hole. A cane laid across the top will simplify the task in that the union point can be levelled with the cane and handfuls of soil placed under the roots to keep the bush at that level; or, more likely, slightly above it because the roots must be firmed in and this will push the plant lower into the hole. This is my method for bush roses and modern climbers; and if standard roses are to be planted, the same procedure of planting would apply, with the soil mark on the stem above the roots being the guide for the depth of planting. In all cases, a good handful of bone meal mixed with the soil in the hole will provide a modest amount of nourishment to give the plant a good start, and in the spring I like to give them all a dressing of fish, blood and bone fertilizer, followed in the latter part of May until mid-August by fortnightly feeds of Maxicrop suitably diluted and watered-in. There are, of course many alternatives, including special mixes of fertilizers formulated for roses.

The necessity for pruning and spraying is regarded by lesser enthusiasts as a substantial reason for not growing roses, but the majority, I am sure, will accept it as a small penalty compared with the enormous pleasure that roses in the garden or cut blooms give. If a winter wash of Jeyes fluid, suitably diluted, is given to the plants and soil, a clean start to the New Year is ensured, and following on from this at regular intervals, based on the advice of the manufacturer, spraying with a combined insecticide and fungicide – and there is now one (Multirose) which also has a 'feeding' agent – should keep matters under control. Unless pruning has been mastered hitherto, reference to a book on the subject is advisable. It is not difficult, however, and largely a matter of common-sense. The centre of a bush would, for example, be kept open by removing completely stems that clutter it up, and if tip-top blooms are sought, hard pruning each year of hybrid teas (large-flowered) should be carried out by cutting all stems back to a few buds above soil level. Living in a warm suburb of London, I find this to be feasible in early March, but slightly earlier is preferred by some; and, of course, in places where the winter harshness lingers, a later date will be necessary. It is normal to be less severe with floribundas (cluster-flowered), but keeping them nice and tidy, and, as with the 'HTs', pruning hard to a few buds above the ground after planting. Modern recurrent flowering climbers are sometimes shy in producing new wood, and it often springs from a point about half-way along an existing stem; so removal of whole stems, or canes as they are sometimes called, has to be on a restricted basis, perhaps removing one or two every other year, depending upon the type of framework that is desired and how many stems grow. The old-fashioned roses which we fondly classified as 'ramblers' produce canes from ground level more readily, and where this is so, canes which have flowered can be completely removed after flowering. In tying canes to supports, it is worth bearing in mind that flowering is induced when the canes are horizontal, but care is necessary to avoid snapping the tender young growths when they are bent to the required shape.

The one feature common to rose blooms wherever grown is that they open quickly after being cut. Bearing in mind that in the perfect stage of development for show purposes a large-flowered bloom should be half to three-quarters open, it must clearly not be approaching the fully open stage when selected for cutting for a show. Practice will indicate how far a bloom should be developed and how quickly it is likely to open out in the interim period between being cut and being judged. Needless to say, there are methods of 'holding back' the opening up of a bloom, and a common practice is to tie the bloom with soft wool, that is, to encircle it and pull the wool comfortably firm, taking care not to tear petals in doing so (figure 10.9). And some exhibitors leave these 'ties' in position until shortly before judging is about to commence.

Figure 10.9
Rose bloom tied
with wool

Roses, as much as any other flower, appreciate a deep drink, and they are probably one of the most suitable candidates for conveying to the show in their containers of water, although reasonable care must be exercised to prevent the blooms bruising each other. Bearing in mind their propensity for opening fast – particularly in warm halls or marquees – cutting earlier than the day before the show is not to be recommended. As with all woody stems, rose stems should be slit at the bottom for 2.5 cm (1 in) or so before being inserted in water. For general convenience, thorns should be removed from the lower half of the stems. If it proves not to be feasible to keep them in their buckets, they can be packed into flower boxes with suitable protection by tissue paper and cotton wool; and the blooms must be protected from damage by damp stems. Or, they can be conveyed in boxes made from plywood which are deep enough to take the length of stem and which have a shelf fixed to the sides of the box and about a third of the way towards the top, and a close-fitting lid, both the shelf and lid having holes through which the stems can be slotted so that the chins of the blooms are clear of the surface of the lid. These can also be protected with wads of cotton wool. Alternatively, a dozen blooms could be wrapped in a bunch and taken by hand, and if this is done carefully there should be no damage caused.

Figure 10.10 Customary styles of arrangement at Rose Shows. (a) Vase of six stems; (b) bowl of nine stems

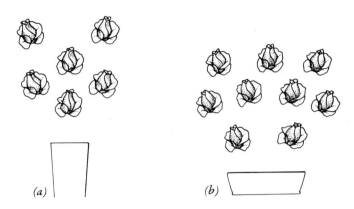

(a) (b)

Staging should not be found to be particularly difficult provided that some packing material is to hand for holding stems in position. Reeds are used, but are not readily available to everyone, and paper is acceptable in vases if it is not allowed to show prominently and preferably is topped by sphagnum moss. Blooms should have slight gaps between them to permit them to be seen individually without overlapping, and convenient arrangements are to set three stems in a triangle, one stem being placed centrally below the other two; and six stems in three rows of two, three and one, from the rear to the front, gently sloping towards the front of the show bench (Figure 10.10, Plates 9-10). Stems in bowls can be inserted in oasis and staged in slightly curving rows to suit the outline of the bowl instead of being arranged to give an 'all-round' effect, although some local societies expect the latter in order that the bowls can be placed in the centre of a table to produce a rather nice effect for visitors. 'Facing the front' in my view, assists the process of viewing with a professional eye, but, as always, the wording of the schedule must be complied with. There are, of course, other styles of staging roses, including placing 'Specimen' blooms in specially constructed boxes which have holes spaced out in a sloping lid and fixed below which are tubes of water into which short stems are placed so that the blooms sit 1.25 cm ($\frac{1}{2}$ in) or so above the top. These boxes of specimen blooms look very attractive, and boxes on similar lines used to stage blooms of miniature roses have a special appeal. A visit to an

'RNRS'-sponsored show will always be worthwhile if only to study the methods of staging concerned – as indeed it will in regard to all other National Society shows – and I think that little better can be done than to follow the practice encountered, and there is nothing ostentatious about doing this in local shows.

One point I must mention, however, concerning local shows is that there are various uses of the terms 'Bloom' and 'Stem'. RNRS guidance is that the former is used only in connection with specimen blooms and three-stage large-flowered roses – an interesting class in which the exhibitor has to show a bud, a bloom in the 'perfect' stage and a fully-open bloom, all of one cultivar. Otherwise, the term 'Stem' is applicable. If the schedule wording is unclear, a quick word with the show secretary will suffice.

On the subject of roses, I should finally point out that miniature cultivars are fast becoming popular. They are a useful alternative for those who cannot plant large bushes but wish to have roses – perhaps in a patio garden, or in a raised bed which can be cultivated from a wheelchair. On the show bench, they look most enchanting, and a substantial number of stems can be coveyed easily to a show. They represent, none the less, a real challenge to the exhibitor because the competition is often keen and a fair amount of skill is demanded to produce a good-looking vase of, say, six stems of either one cultivar or of more than one cultivar. I can thoroughly recommend that exhibiting miniature roses be seriously considered as a challenging and interesting subject.

Shrubs and Trees

I do not think that it is practical to select a list of shrubs and trees for exhibiting; because of their size they must lend themselves to the general attractiveness of the garden and will need to be 'in scale' and so forth. Shrubs, and even more so trees, are of a dominating character and usually fairly permanent residents, so great care in choosing what is suitable is always to be advised. They are, however, strong candidates for showing in spring, summer and autumn, and thus, it might be said, can add another string to one's bow. If one is fortunate enough to acquire a well-stocked garden containing shrubs and trees, then hopefully there will be something among them that offers itself as potentially good exhibition material, whether it be as 'Foliage', a 'Berried' shrub or a 'Flowering' one. Care must be taken in reading the show schedule on these points to determine exactly what is required: if, in fact it asks for an exhibit of 'Berried' shrub, then obviously plenty of berries should be present on the stems that are selected. It is of great importance, too, to make sure that the berries look fresh and are not drying-off.

Similarly, stems cut from a 'Flowering' shrub should carry lots of open flowers and not a predominance of buds. These points seem elementary in their nature, but it is surprising how many exhibitors fail to observe them and suffer disappointment thereby.

Stems should be cut at the latest possible time. Woody stems must be slit with a knife or by gentle hammering which should not crush the stem ends into a pulpy mess, and a good long drink of water is required. Dipping stems of such shrubs as lilac (syringa) into boiling water can help to prevent a callus forming, but care must be taken to protect flower heads from steam; and stems which exude latex should have the ends cauterised by candle flame or a lighted match. Very often, a schedule will not specify a required number of stems, but equally often certain dimensions are stated – for example, height and width or breadth of 'two feet, six inches' (some 75 cm). These measurements must not be exceeded, and where necessary, stems and foliage must be trimmed accordingly. A common mistake is to overcrowd a vase, and far better it is to present three stems well clothed in foliage, flowers or berries, nicely arranged so that each stem is seen as an entity, than to jam six or seven stems of varying qualities into the vase making it difficult for the judges not to take the view that it all looks rather a mess. A good length of stem should be shown above the top of the vase, within any stated maximum height in the schedule, and this can be done by lifting it up with paper and sphagnum moss placed at the bottom of the vase, but remembering also to wedge the stems firmly into position to guard against their falling about if and when the judges lift the vases onto the floor for a better look at the exhibit, or into the air so that any signs of damage underneath the foliage can be spotted.

One of the easiest means of conveying material cut from shrubs or trees is to place it in a deep flower box, and to carry that, without a lid, on the rear seat of a car, but if it has to be packed for carrying some distance on foot and therefore the lid has to be fitted, then clearly only a very deep box will be of any service, and crushing must be avoided. This will probably entail manufacturing a box from two boxes joined together with straps, the one on top with the bottom removed. It all depends upon the depth of the material.

With individual blooms, camellias are an obvious choice for a spring show. It is usual to exhibit these with two leaves attached, the short stem sitting in a small vase or pot so that the bloom tilts slightly towards the front. It is well known that camellias are not tolerant of alkaline conditions, and I find that an annual dose of a proprietary sequestered iron product is necessary to produce glossy green leaves. But, if room can be found for one or two plants, it is well worth coping with the relatively small amount of bother – perhaps by digging out a substantial hole and filling it with an acid peat mixture, and applying annual doses of Sequestrene – because of the joy of seeing the lovely, exotic-looking blooms in spring.

Bulbs

Bulbs can be grown successfully in pots and other containers, and many societies have a class in their spring schedule for a container of daffodils, which, apart from providing a colourful exhibit, enables those without a garden or physically unable to cope with ordinary gardening requirements to enter the competition. Four bulbs planted in a 22.5 cm (9 in) pot containing compost will produce a good display of eight or more stems of flowers. The bulbs should be planted in September about a third of the way down the pot, to set their noses at about surface level. They then need a cool environment, and in a garden this can be achieved by burying the pots, with 7.5 cm (3 in) of soil covering them. If a really hard spell of frost is forecast, additional covering with several inches of bracken or peat will be necessary. For flat-dwellers, a balcony or a spot by a porch will need to be found. It is important not to allow the compost to dry out, but, of course, good drainage is essential, and as is customary, the pot should be 'crocked' by placing some pieces of broken clay flowerpot or similar material in the bottom. Attention to watering is more likely to be needed where pots are not buried in the garden, and in such cases it will be necessary to provide protection from hard frost by tying sacking and thick newspaper over and round the pot or bringing it inside the porch. The matter of judging when to bring the pot 'inside', whether that be a greenhouse or a kitchen area, will depend upon the weather, and primarily the temperature. In sheltered areas and where some heat is to be provided, early March would be acceptable for producing blooms for a spring show, but there is scope for holding blooms back by keeping the bulbs in a cool place or forcing them along in a warm one; and it is by no means unusual to use the 'living-room' for this purpose. Strong growth, however, is best achieved in a cool spot where the light is good.

As with every other aspect of horticultural science, practice and experience are invaluable aids in support of theoretical knowledge, and it will take a year or two to determine the best method. It really is surprising to see the wide range of techniques employed: some exhibitors, for example, show the same pot of bulbs for several years running, probably ensuring sufficient initial nourishment each year by replacing the top few inches of compost with

Daffodils

fresh material mixed with a small handful of bone meal. Others may prefer to buy each year plump single-nosed bulbs, thus virtually guaranteeing good results. One cultivar that seems to cheerfully accept the 'several years running' technique is 'Mount Hood' which produces sturdy stems and nice-sized white blooms that last very well. These are helped along if, when flower buds appear, liquid feeds of Maxicrop or Phostrogen, or perhaps alternative doses of one of those and a high potash feeder such as Compure K are given at fortnightly intervals.

Experienced exhibitors also use the pot-growing method to ensure that they have blooms ready for a show because they can accelerate or retard growth in the controlled environment of a greenhouse, taking the pots in and out as needs be: or, the 'spare' room indoors can be used, with radiators being turned on or off as necessary. The same degree of control is not available in the case of open-ground cultivation, and blooms are subject to weather damage. None the less many good blooms are produced from outdoor growing, and they can be protected from being damaged by wind and rain by being cut when the colour is showing in the buds – preferably when they have set themselves at the appropriate angle to the stem – in most cases 90°. The length of time for development into a fully open bloom will depend upon temperature ranges, which can, of course, be controlled indoors. The stems must remain throughout in containers of water, and if the bottoms become sealed in any way, a small piece must be cut off and the stems replaced in water immediately. Refrigeration techniques are another matter and, I think, for individual experimentation, if so desired.

A satisfactory method of growing daffodils in the open is to have them in beds of a width of 1.5 m (5 ft) so that cultivation is possible from paths on either side. The bulbs should be planted 15 cm (6 in) apart in rows 30 cm (1 ft) apart. Daffodils, and many other bulbs, including tulips, are not especially demanding in their soil requirements, but plenty of humus produced from digging in composted organic material is highly desirable; and a sprinkling of bone meal at planting time with liquid feeds when the buds appear, given at fortnightly intervals, will help. When it comes to cutting stems for the show, buckets should be to hand so that so soon as the stems have been severed by a slanting cut with a very sharp knife they can be plunged into water. Containers with squared mesh wire halfway up and again at the top can be used not only to hold the stems whilst they have their deep drink, but also for conveying them to the show; but there are adequate ways of packing them into flower boxes. This involves protecting the blooms from coming into contact with damp stems, and greaseproof paper can be used as a barrier. If the box is first lined with tissue paper or cotton wool, the stems can be laid in rows running the length of the box, with the blooms just under each row of blooms above and resting on a wad of cotton wool placed on top of the greaseproof paper covering the stems. Quite a sophisticated arrangement can be made simply by rolling newspaper very tightly into sticks or rods of about 2 cm ($\frac{3}{4}$ in) in diameter, and rolling these in tissue paper. They are then packed tightly lengthwise into the box so that stems can be trapped between them without bruising, and the blooms sit comfortably on top of the tissue paper. In this way, they do not touch the stems, and an additional safeguard would be to pad the blooms underneath with cotton wool. When the stems are taken from the boxes, if there is any sign of sealing over at the bottom, a fresh cut should be made before inserting the stems into their vases.

In staging the blooms, the stems need to have a good stiff poise, so they will have to be firm in the vase, and reeds or sphagnum moss or paper topped with the latter will be needed. Each bloom should be seen in its complete form and not in any way hidden from immediate view, but neither should there be large gaps between them. Leaves add a pleasing touch, and they should be placed with their tips coming about 2.5 cm (1 in) below the blooms. As a general guide, two or three leaves would be appropriate for a single stem, and four or five for three stems. The leaves are intended to enhance the appearance of the exhibit and not to outweigh the flowers. Three stems should be set in a triangular arrangement, with one bloom at the apex, and apart from cultivars that naturally hang their heads somewhat, the exhibitor should do his or her best to make the blooms look the judge in the eye. They may not have acquired the desirable set of 90° to the stem, and in that case discreet tilting back of the stem may do the trick, although the judge will note this point. Desirably, the perianth (petal segments) of the trumpet daffodils and the long- and short-cupped cultivars should be flat, and the top and bottom segments should point to twelve and six o'clock, respectively (Figure 10.11). Needless to say, experienced exhibitors find it fairly easy to persuade blooms to adopt the right poise if they have not grown in the desired manner. Petals are coaxed into a flat position by firmish but careful use of a camel hair brush and the whole bloom is turned by grasping it gently but firmly at the back and, holding the stem at the top, twisting the bloom in the required direction. Obviously this is something that requires practice, preferably with blooms that are not required for the show.

One difficulty about daffodil exhibiting at the moment is that societies vary in their practice of classifying them. The

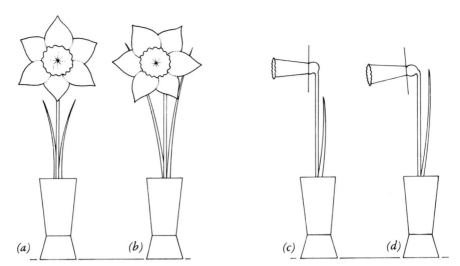

(a)   (b)   (c)   (d)

Figure 10.11
Top and bottom
petals of daffodil
should be in vertical
line (a) and not
'clocked' (b);
trumpet should be
at a right-angle
to the stem (c) and
not drooping (d)

Daffodil Society uses the name 'Daffodil' as the accepted term for embracing all the flowers that in the past many of us were accustomed to dividing into daffodils (those with large trumpets) and narcissi (those without trumpet). So 'Daffodil' it now is; and societies are slowly moving towards the new terminology which classifies daffodils into twelve divisions.

Division 1 contains the trumpet daffodils, and the trumpet has to be at least as long as the length of the perianth segments (petals).

Division 2 is for long-cupped cultivars, with the length of corona (cup) being more than one-third of, but less than equal to, the length of the perianth segments.

Division 3 embraces the short-cupped cultivars; that is, the length of cup must be no more than one-third of the length of the perianth segments.

Division 4 is for double daffodils; that is, cultivars which produce double flowers.

There follow eight more divisions covering all other types of daffodils that are encountered; *Triandrus* (with the bowed head), *Cyclamineus, Jonquilla, Tazetta* (including the well-known cultivar 'Geranium' with its lovely scent), *Poeticus* (including 'Pheasants Eye' and not to be confused with a short-cupped (Div. 3) cultivar), 'Species and Wild forms', 'Split-corona' cultivars and 'Miscellaneous' covering anything not already specified. The Daffodil Society Handbook explains in more detail what is involved and many societies print explanatory notes in their schedules.

There is also a format for colour coding each cultivar using the initials 'W' for white or whitish, 'G' for green, 'Y' for yellow, 'P' for pink, 'O' for orange and 'R' for red. Blooms are categorised by division followed by their colour description, starting with the perianth and followed by the cup or corona which for colour purposes is divided into an eye-zone, a mid-zone and an edge or rim, the colour code letters being placed in that order. This may seem complicated but if we take as an example the Division 9 cultivar 'Pheasants Eye', the classification can be determined as '9W-GYR': that is, it has a white perianth and a cup (corona) starting with green in the centre or eye, followed by yellow and then red at the edge of the cup. 'Mount Hood' demands less because it is simply all-white and is therefore described as '1W-W'; quite a neat and brief way of describing an all-white trumpet daffodil, the hyphen separating the perianth and the trumpet.

It is not common practice among societies to have classes for all divisions of daffodils or to call for a multitude of colour ranges. Very often, however, there will be separate classes for all-yellow and all-white trumpet daffodils, that is 1Y-Y and 1W-W, because several cultivars in each category are widely grown and therefore a goodly number of entries in each of those classes can be expected. In the former class there are, for example, 'King Alfred', 'Rembrandt' and 'Unsurpassable', and in the all-white bracket we find 'Beersheba', 'Empress of Ireland' and 'Mount Hood'. Whilst some of the old favourites such as 'King Alfred' and 'Mount Hood' are still widely grown and shown in local shows, there are in all divisions cultivars that are known for their special qualities on the show bench, one important feature being a flat perianth. As with most things, to get what is considered the very best requires a fairly substantial financial outlay, depending, of course upon the numbers of bulbs that are to be purchased. If it is wished to purchase a modest number to begin with and to pay a price that is not at the top of the range but will guarantee good show blooms – always given that cultivation is carried out diligently – it could pay to buy the cultivars 'Kingscourt' and 'Lemon Meringue' for exhibiting in the trumpet division. They are both all-yellow (1Y-Y). For a large-cupped entry, it would be worth purchasing 'Desdemona', a lovely all-white cultivar, and 'Tudor Minstrel' which has a white perianth and a yellow corona. Good short-cupped cultivars are 'Woodland Star' (white perianth and orange/red corona) and 'Verona' (all-white). Double cultivars worth considering are 'Papua' (all-yellow) and 'Tahiti' (yellow perianth, yellow, red, red cup (4Y-YRR)). These are but a few suggestions for a start, involving cultivars that will produce good show blooms without 'breaking the bank'. Undoubtedly it would be to one's advantage to extend the range beyond the four divisions I have mentioned, provided, that is, that the show schedule is suitably comprehensive. Two that therefore could be added would be a nice *Cyclamineus* cultivar classified as 6W-Y and named 'Dove Wings', and the well-known cultivar 'Geranium', already mentioned above, and classified as 8W-O.

Tulips and Hyacinths

In my experience, daffodils represent the major part of spring shows, but there are often attractive possibilities concerning other bulbs, such as a class for a bowl of species tulips or grape hyacinths, and these offer the flat-dweller the chance of competing on equal terms; perhaps of most significance in this regard is 'a bowl of Hyacinths' where the trusted technique of starting the bulbs off in a dark place and gradually bringing them into the light can be practised with little hassle. If cut tulip blooms are to be shown, there are one or two points to consider. First, the blooms will open very quickly so soon as they come into a warm atmosphere, so everything must be done to keep them cool. Secondly, the stems tend to twist out of a straight line; and this can be corrected by wrapping them firmly to just below the flower heads in damp paper and gently stroking them into an untwisted posture.

Lilies

So much for spring: for summer shows there are lilies, and some of them are easily grown in the border or in a pot. The 'Regal Lily' (*Lilium regale*) is a good choice because although not the most exotic of lilies, it performs well in a wide range of soils; but not, of course in waterlogged conditions, so good drainage is necessary. If it can be contrived, a position where the lily can have its head in the sun and its lower stem shaded should be beneficial. The herbaceous border is a suitable place; and the bulbs should be planted about 20 cm (8 in) deep with some sharp sand at the bottom of the hole, in the early autumn. The 'Madonna Lily' (*Lilium candidum*) is also a fairly tolerant plant, but it does not require deep planting; barely 2.5 cm (1 in) of soil should cover it. To overcome the consequent problem of winter damage, several inches of peat can be used to provide protection, and this can be removed when the shoots appear in the spring. The 'Tiger Lily' (*Lilium tigrinum*) is yet another good choice because of its general benevolence towards the gardener, and this requires planting in the manner described for the Regal Lily. A lot of very good lilies are grown in pots of 17.5-20 cm (7-8 in) diameter, and these need to be well-crocked and filled with good loam and composted organic material or a good potting compost, with a little rotted manure at the bottom; some compromise about planting depth can be made because the pots can be 'housed' for the winter. Stem-rooting cultivars, such as the Regal lily should, however, be planted well below the surface.

Corms and Tubers

Coming to the tubers and corms, it is probably worthwhile planting some anemones of the St Brigid type, and if it is wished to raise the small tubers (known generally as corms) from seed this can be done from a spring sowing, but the seed is notably slow to germinate. Planted in spring in a sunny border, the tubers will produce flowers in summer, and planted later on – say, in September, they should produce flowers in the spring, always depending upon the temperature range. Certainly no great effort is involved in planting them a few inches apart in loamy soil, at a depth of 5 cm (2 in), and a vase of the delightful blooms will add beauty to the show bench.

Anemones

But it is to the gladiolus that I would direct the reader's attention because it is a majestic flower that can be shown in summer shows and, by staggering the planting of the corms over several weeks, it is possible to have blooms for September shows as well. The corms should be planted at a depth of about 10 cm (4 in) from mid-March (in areas with comparatively mild weather) until the end of April or slightly later. It is customary to place a good sprinkling of sharp sand in the base of the planting hole to make the corm feel more comfortable than sitting on wet soil. The plants will appreciate facing the sun, but the soil should have sufficient water-retaining capability to prevent its drying out frequently. Hopefully, therefore, it will be found to be possible to dig-in a good measure of material from the compost heap, supplemented as needs be by peat. When the flower spike is seen to be emerging between the leaves, liquid feeds can be administered at ten-day intervals; Maxicrop or Phostrogen will be suitable for this purpose. Staking will be necessary.

Gladiolus

There are miniature, small, medium, large and giant cultivars of gladioli, with a range of flower diameter from under 6 cm (about $2\frac{1}{2}$ in) to more than 14 cm ($5\frac{1}{2}$ in). The British Gladiolus Society has produced a classification list of size and colour of flowers of the many cultivars that can be grown, and membership of the Society would undoubtedly be of benefit to anyone taking a keen interest in gladioli. Local shows often have various size parameters for gladioli, and I think that it would normally serve well to buy corms of medium-size cultivars which should produce flowers within a range of more than 8.75 to 11.25 cm ($3\frac{1}{2}$-$4\frac{1}{2}$ in) in diameter, although there should be no penalty if they were to

Figure 10.12
A good-quality
gladiolus stem

exceed that and fall into the 'large' size range of about 11.5-13 cm (4-5 in).

The spike required for a good-quality exhibit of a medium-size cultivar should have a flower head of at least 50 cm (20 in) in length, with an optimum seven open flowers plus five buds showing colour, topped by six green buds; and it should be perfectly straight with no curly tip (Figure 10.12). A 'tall order', I think, and there is no cause to feel ashamed of spikes with somewhat less blooms and buds, the main thing being the production of an exhibit that looks balanced and attractive. Thus, a spike with five or six flowers fully-opened with just one or two buds on top would be less desirable than one with four flowers opened, two half-opened and two buds showing colour, plus two green buds at the very top. Technically, too, the half-opened flowers would rate as 'open flowers' and not as buds. And as regards keeping a straight tip to the spike, judicious tying to a stake or thick cane should help.

Spikes can be cut as soon as the bottom bud is seen to be opening or at any time thereafter, provided that the bottom flowers have not reached a stage of having been fully open for a day or so, because by then it is very likely that at least one of them will be drooping by the time of judging should the hall or tent be hot. If the weather is kind enough to permit cutting on the day before the show, the spike at that stage should have bottom flowers opening-up to full size and looking fresh, with some more flowers being up to three-quarters open and perhaps one or two about half-open; and with some buds showing colour. If it is known that some bad weather is brewing, it would obviously be discreet to cut earlier in the week before the show, and the flowers would then need to be at a less advanced stage. Further development can be influenced by the conditions in which the spikes are kept, but, as with most kinds of flowers, one would not want them in a hot, dry atmosphere, and they must stand in water throughout.

An important point to watch for is that the tips of the spikes will curl towards any distant light source, so placing them in their containers some distance from a window is something to be avoided. When they are taken out of their containers for packing into flower boxes – if that is the method by which they are to be conveyed to the show – the stems should be wiped with a cloth to remove excess water; and the flowers will need to be well protected from crushing. Tissue paper and cotton wool will be required, the latter being useful for padding underneath the blooms. Notches or holes in one end of the box will probably be necessary to accommodate the ends of long stems.

Staging the spikes should aim at presenting them in as

(a)   (b)

Figure 10.13
Flower spikes
should be upright
(a) and not with
curled tip (b)

upright a position as possible, and they must be firmly wedged in their tall vases (Figure 10.13, Plate 5); newspaper and sphagnum moss can be used for this purpose, with the paper being hidden from view. If, say, three spikes are to be exhibited in one vase, a little fanning out will be necessary so that the blooms of each spike are separated – but without exaggerated leaning to left or right. There is a lot that can be learned from watching exhibitors at major shows, and from specialist books, about 'Dressing' blooms. I have seen some persuasive manipulation carried out; but, until one has had a chance to practice it, I think it would be best to leave the blooms untouched. One thing that can be done, however, is to remove a bottom floret, or even two (but no more) if distress and tiredness are evident. This will result in 'downpointing', but it is better to do this than exhibit a spike with dying blooms.

Gladioli corms obligingly manufacture a new corm each year to take the place of the old one which is discarded. They will only produce a satisfactory new corm, however, if, when the flower spikes are cut, some undamaged leaves are left to nourish the production process. So, whilst the spike should be cut with a long 'handle', care must be taken not to sever all the leaves. In due course, after the plant is lifted, the leaves will be cut off close to the corm which, with its old corm attached underneath, will be dried in an airy place. The

*Figure 10.14
Separating a new
gladiolus corm from
the old growth*

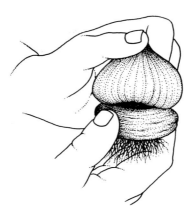

old shrivelled corm can be removed when it is found to be easy to separate without a great deal of effort (Figure 10.14), and the new corm stored in a cool but frostproof place until the time comes for planting it and repeating the cycle.

Gladioli are well worth the little bother that seems to be involved, and if it is thought to be too difficult to cope with the larger spikes, particularly perhaps in conveying them to the show, there is an alternative of growing the delightful *Primulinus* type. This is by no means a second best, and the skill of cultivation remains: and on the show bench they look intriguingly beautiful.

11. A Selection of Fruit

Some fruits cannot be said to be reliable croppers in the United Kingdom, unless they are grown in sheltered spots or under cover; and to purchase with a view to exhibiting without having the right environment would amount to a gamble. Apricots, for example, can be grown in large pots if appropriate treatment is given to restrict the trees to a small stature, but in the open they have to cope with chilly conditions when the blossom is ready to set – at least, I find this often to be so, and consequently that a poor set results. Similarly, the small fruits forming on a fig tree will capitulate to frost in the spring unless some covering is provided. Both these fruits are, however, likely to win approval at a show if they can be satisfactorily grown; and whilst I seldom see apricots, I often encounter dishes of ripe figs, so there are obviously plenty of gardeners who know what is required to master the art of growing them.

Individual Fruits

Apples

Apples should present less problems, although if one is blessed with a large standard tree, it is no mean feat to prune it and to gather the crop. If starting from scratch, dwarf pyramid or cordon trees should be purchased. This will bring the effort of looking after them within most people's reach, and a dwarf pyramid can be satisfactorily grown in a large pot, provided that it is watered copiously in hot, dry weather. Nurserymen will advise on the most suitable cultivars for growing in pots and also on the type of dwarfing rootstock that one should choose to achieve particular aims. If exhibiting is the main aim rather than weight of crop, it helps to grow several trees in a small area provided that they have been grafted onto a rootstock that ensures a dwarfing habit of growth. Pruning and spraying are easily carried out – the size of the task involved in pruning and spraying a large tree is often so daunting that it militates against its being done at all. I think that unless the art of pruning has been already mastered, a textbook should be consulted; or, better still, go to a demonstration of the art

– to the RHS Garden at Wisley, if possible. Most people shape their trees nicely, and an attractive tree is desirable, of course. It is also important to induce fruit spurs – the stubby looking shoots that will produce flowers and fruit (Figure 11.1) – and to ensure that sunlight can filter through and add some colour to the fruits.

Figure 11.1
A fruit spur

Spraying is demanding of some time, but necessary, unless some 'safe' alternative by way perhaps of biological control is feasible. A 'winter wash' of tar oil or similar product will clean away the lichen from the trunk and branches and pave the way for a healthy season of growth, but seldom is that sufficient, and a product such as Fenitrothion should be used in the spring and early summer. It is vitally important to follow the maker's instructions when using any of these products, both in timing and weather conditions, and also in handling them safely. Pets, birds, neighbours must all be thought about before launching out on the spray crusade, and with a little consideration for them, it should be possible to select a time when the only offence committed will be against the unfriendly insects that needs must be destroyed. As if spraying against insects were not enough, there is also a requirement to prevent fungal complaints such as mildew, and here there are some fairly 'safe' products on sale generally, and in certain cases it is possible to combine an insecticide with a fungicide, if specified by the manufacturers.

We all have lots of 'windfalls' every year and this combined with the 'June drop' when masses of small fruits tumble from the trees, helps to reduce the crop to a volume that the tree can cope with. But experienced exhibitors carry out a thinning programme over several weeks by removing misshapen fruits and those that would impede the growth of others; and twiggy wood which might perforate fruits of quality is pruned away. A further step is to enclose chosen fruits in muslin bags to prevent bird damage. All this takes time, and if one is just embarking on exhibiting, it would be more advisable to take some of the measures I have mentioned on a modest scale; a little thinning can be done, for example, rather than attempting to turn every apple into a potential winner by rigorously discarding all doubtfully shaped specimens. It could all become 'too much' for a start, but more acceptable when experience has been gained. So far as bird damage is concerned, one advantage of growing dwarf pyramids is that a lofty fruit cage can be constructed to enclose them.

One thing that can be done without any real bother is to nourish the trees. It is not at all uncommon for everything in the garden to be treated annually to a feed of a general fertilizer barring the poor old apple tree which is expected to perform miracles each year on a starvation diet. So when the Growmore or fish, blood and bone is being distributed to the plants in spring, please scatter a couple of handfuls round each apple tree. In the early summer, 30-60 g (1-2 oz) of sulphate of potash sprinkled onto the surface and watered-in will help to bring colour into the fruits.

When selecting fruits for showing, uniformity of size, shape and colour should be sought. When selecting for size, it should be borne in mind that a dessert apple ought to look suitable for holding in the hand for eating; therefore very large specimens are not required, and a desirable maximum diamter is 7 cm (2¾ in). It is stated in the RHS Handbook, however, that certain cultivars are not to be regarded as faulty if they somewhat exceed that size. 'Blenheim Orange' and 'Charles Ross' are in this category: and as they are also eligible to be shown as cookers, in that circumstance they would need to be as big as possible!

Remembering that the whole stalk must be retained, it would be best to practise harvesting fruits to determine the best way of picking them without pulling the stalk away from the fruit. When ripe, the apples should part easily from the tree if a hand is placed underneath and a slight upward twist given. If that is a frightening thought, lest the stalk might become detached, the use of a sharp pair of long-bladed scissors might be better, taking care to secure all the stalk and not just part of it. Personally, I would wear thin cotton gloves to minimise marking the 'bloom' on the apples, some cultivars of which have that lovely natural sheen which ought to be preserved. If the fruits are placed straight away into tissue paper-lined boxes, they need not be handled again until staging time at the show. It is not necessary, as with many other items for the show, to wait until the day before show day to pick apples. Most cultivars will keep for several days at least without showing signs of lack of freshness, and with one or two cultivars it is even possible to pick them and lay them *gently* on the lawn to colour up in the sunshine over a few days. But the risk from an inquisitive bird or animal is there, and must be guarded against. Apples should be stored in a dark cool place, and for very late shows they will have to keep their best looks for quite a while, so the atmosphere must not become dry.

At the show, the fruits should be taken carefully from their containers by holding the stalks, but with fingertips taking the weight underneath. Apples should be staged on a plate, the well of which can be filled with tissue paper (white and not coloured); and a further piece can be enveloped round the plate. This is not essential at local shows, but conforms with practice commonly adopted at major shows. The important thing is to use a white plate rather than place the fruits onto the covering on the bench, and this little attention to detail is worthwhile because the exhibit is enhanced in appearance. A simple and pleasing arrangement is all that is required, and fancy styles should not be entertained (see Plate 17). The fruits should be placed stalk downwards, and three fruits should form an equilateral triangle. Five or more can be staged on the basis that one is placed in the centre of a circle of the others, the centre fruit being raised a little on a wad of tissue paper.

Apples are usually very much in prominence at September shows and are also a useful item for a 'Late Autumn' show. As I have already mentioned, small trees can be cultivated, and this widens the possibility of being able to grow and show apples. As well as seeing a number of dwarf pyramids in modest-sized gardens, I have seen a whole run of cordons at the rear of a flower border, and provided that the flowers do not encroach upon the trees or poach their food reserves, this seems a reasonable arrangement, particularly as the blossom in springtime is so delightful. As the trees will stay in their positions for a long time, it follows that they should be planted in good soil with plenty of humus content. Compost must be added if the soil looks 'thin', and free drainage of clayish soil must be engineered by at least working into the soil some coarse, sharp sand and peat, in proportion of about 'one to three'. A good handful of bone meal per tree at planting time will not come amiss. But I am sure that in the case of anyone in doubt about how to plant new trees, the nurseryman will be happy to explain everything.

Blackcurrants

Just one blackcurrant bush, well-grown, will yield enough fruit to enable several pounds of jam to be made, a jar of which can be entered in the Domestic section, and will also provide attractive specimens for a class in the Fruit section. Certainly the large, jet-black fruits can look handsome on the show bench, and there is no great difficulty in producing them if there is space for a bush in the garden or on the allotment. Management of the bushes is aided by the fact that some of the pruning at least can be done in conjunction with picking the fruit in the summer. The principle is to remove as much of the old wood as possible, and the centre of the bush should be kept free from being cluttered up with criss-crossing branches, so that light and air can enter freely. The new wood is easily distinguishable because of its light colour in comparison with the older, fruited wood which will have ripened into a deeper brown shade: hopefully some new shoots will spring each year from the bottom of the bush, near ground level, and this will enable older stems to be completely removed. When new bushes are planted in the autumn, all the shoots should be cut back to not much more than 2.5 cm (1 in) from the ground. Strong shoots for fruiting in the second summer after planting can thus be expected to develop, and really it is best to remove embryo fruits that appear in the first summer, but I realise just how difficult it is to discard the possibility of tasting at least a few fruits as soon as possible!

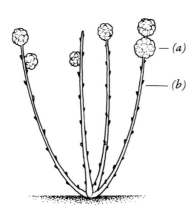

Figure 11.2
(a) 'Big bud' on blackcurrant bush. These must be picked off and burned. (b) Normal bud

If, instead of nicely pointed buds, some plump, roundish ones appear, 'Big Bud' will have manifested itself, and it must be dealt with immediately by removing and burning the offending buds which have a mite inside them (Figure 11.2).

Sometimes aphids make an appearance in the late spring, and spraying the stems and leaves of the bush thoroughly with Malathion just before the flower buds open should deal with the problem. The other 'danger' is that birds may take a liking to ripe fruits, so netting the bushes will then be necessary; but I do not find that birds are a nuisance concerning blackcurrants, although they always attack my redcurrants unless they are protected by netting. New bushes can be propagated easily from cuttings of healthy young shoots, partly ripened, complete with buds, and about 20 cm (8 in) long. They should be planted about 12 cms (5 in) deep in a slit in the soil with a spade. A sprinkling of sharp sand in the bottom will help, and the cuttings should be made firm. Planted thus in the autumn, they should, by the following autumn, have a good root system, and they can then be planted in their permanent site. It is interesting to propagate from cuttings, but on no account should they be taken from dubious-looking stock: it would be far better to invest the modest sum required in clean stock from a nursery.

The freshness of the fruit is an important factor in whether it will win an award, and the strigs should not, if possible, be harvested earlier than the day before the show. They will need to be cut carefully from the branches so that the full length of stalk is secured, and nicely tapering strigs with the appearance of a well-formed, if small, bunch of grapes, should be chosen. Placed in tissue paper-lined containers and taken into a cool place, they can be conveyed to the show without further handling. At the show venue, they can be laid carefully onto tissue paper on a staging table for final selection. They should be staged in lines across a plate with the bottoms of strigs towards the front of the bench. Sometimes they are mounded slightly onto a wad of tissue paper, and if that accords with a local custom, it would seem reasonable to follow that practice. Whatever the layout, the exhibit should look neat and attractive.

Gooseberries

Of all the soft fruits, I know of none that is so neglected yet expected to respond year upon year with generous crops. The difficulty that arises if bushes are not pruned to make them less than a mass of intertwining branches, is that picking becomes a savage business, and the soreness of one's forearms is only partly compensated by the delicious flavour of home-made gooseberry tart. Bearing in mind that for show purposes the stalks must be retained, cutting the fruits from the bush with sharp scissors is a necessity unless one has a particularly long and sharp thumbnail, and even then undue pressure might be exerted upon the fruit, so bruising it. So the fruit needs to be fairly accessible to enable a pair of scissors to be used without difficulty; and keeping the centre of the bush open should be a matter of annual routine. This will also enable sunshine to penetrate and so help the fruits to ripen. An alternative to the bush is to grow on a cordon system, and then picking will be a simple matter, and, of course, several cultivars could be grown in a relatively small area. There is often a separate class at shows for the splendid dessert cultivar 'Leveller' whose size is, or can be, quite dramatic. Because many cultivars of gooseberries are not necessarily easily obtainable except from a specialist nursery, I do not often see a wide range of them at local shows. If possible, however, I would include a red cultivar in a selection of a few bushes, and 'Whinham's Industry' is a reliable cropper in this regard.

A good mulch each year with compost and two handfuls of fish, blood and bone fertilizer scratched into the soil around each well-established bush will, if applied in the spring, help to produce a healthy crop. But American Gooseberry Mildew must be thwarted by precautionary spraying with Benomyl or a similar product as the flowers begin to open and later at two fortnightly intervals when the fruitlets are forming. It is of paramount importance to follow the producer's instructions, and spraying when blooms are open and when bees are around is taboo as far as I am concerned. The other precaution that is required is to take action against birds if they show interest in the fruits, and this seems to depend upon the environment in which the bushes are grown; some are more fortunate than others in this respect and find that birds do relatively little damage.

If harvesting can be delayed until the day before the show, this would be the optimum timing, but berries will usually keep fresh for a while longer in a cool place. Although they look robust, they should be handled by their stalks with care and when staged they should be set out on a plate in lines, with the stalks facing away from the front of the bench.

Raspberries

I consider that planting a few raspberry canes represents a good investment, because a crop of delicious fruit is practically guaranteed each year provided that some fairly basic steps are taken to cultivate the plants. Early November is a good time to plant new canes which should have their roots covered with 7.5-10 cm (3-4 in) of soil (Figure 11.3). A high humus content produced from composted organic material is desirable and it is worth pandering to the plants' demands for potash by sprinkling-in sulphate of potash – say, 15 g ($\frac{1}{2}$ oz) per cane. Canes should be cut down to some 22.5 cm (9 in) soon after planting, and if they are not allowed to fruit in their first season of growth they will be all the stronger, although autumn-fruiting cultivars can be allowed to bear a little fruit. Annual maintenance involves pruning away the old fruited wood of summer-fruiting cultivars after fruiting and of autumn-fruiting cultivars at the end of February. The new canes growing from the ground need to be tied to supporting wires secured to some strong stakes: a customary method is to have two parallel wires stretched between the stakes or posts at heights of 1 m and 1.5 m (3 ft and 5 ft) and at an appropriate time during the growing season, the canes should be shortened so that their tips are about 15 cm (6 in) above the top wire. An annual mulch in May accompanied by a sprinkling of sulphate of potash at about 25 g (¾ oz) per metre run should help to keep the plants growing lustily. The most feared trouble of infestation by raspberry beetle can be dealt with by spraying or dusting with derris when the young fruitlets are setting; not, of course when the flowers are fully open and the bees are around. Netting against attack from birds will probably be necessary.

Figure 11.3
Planting a raspberry
cane. (a) Too deep;
(b) correctly

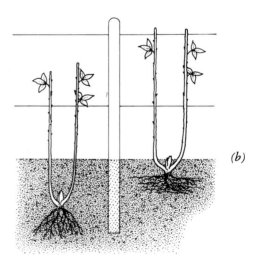

(b)

The fruits should be snipped from the canes with stalks intact and placed carefully into a tissue paper-lined container, and then kept in the cool. They tend to lose their brightness fairly quickly, so only one night in storage is the desirable maximum; and picking any fruits that look to be very ripe indeed should be avoided. At the show, they can be laid out onto tissue paper for final selection, the berries being chosen for their uniformity of colour, size and freshness, and handled carefully by their stalks. They should be placed on a plate in lines across, stalks to the rear of the bench. Size varies according to cultivar and there are several good large-fruiting cultivars from which to choose and details of which are freely provided by nurserymen.

Strawberries

Many of us have fond memories of the superb exhibit of 'Royal Sovereign' strawberries shown annually by the Waterperry School at Chelsea, and for classical shape the cultivar remains a favourite. But, of course, there are many varieties from which to choose, and apart from differences in size and shape, some have more disease tolerance than others, a factor which I think should come high on the list when buying plants. As with other fruits, an abundance of information is usually available from nurserymen and in garden centres concerning the virtues of different cultivars and their fruiting times. It would be wise to plant two or three cultivars to cover the summer period and also to produce some fruits in September, and this is facilitated by the fact that some cultivars fruit in some quantity twice a year.

If plants are set out in the ground or into pots in August, a good crop can be expected the following year, and this will be helped if the soil has a high humus content. Bone meal or fish, blood and bone sprinkled round the plants and lightly raked-in at a rate of about a small handful per metre (yd) run or about 30 g (1 oz) in a pot of 20 cm (8 in) diameter will help to establish strong plants. Planting should leave the crown of the plant level with the soil, but, of course, there is the alternative of planting strawberries in purpose-made barrels which are supplied complete with instructions. In the open ground, the fruitlets will need protection from contamination with the soil, and this can be done by tucking straw or specially-made 'mats' under them. They will also need protection from birds and slugs, so netting and a few slug pellets will probably have to be used. Too close planting should be avoided so that it is easy to work between the plants and rows when clearing away all the debris of spent foliage, straw and the like at the end of season. Something of the order of 30 cm (12 in) between plants and 45 cm (18 in) between rows will provide ample working space, but it is a matter of some compromise in many cases so that the maximum number of plants are grown in a comparatively small area.

However, plants should not be expected to serve well after, say, three years of fruiting, and there are some people who plant afresh every year. New plants can be propagated in the second growing season of the original plants by pegging into small pots – sunk in the ground beside the 'mother' plant – some of the trailing shoots, known as 'runners', below their growing tips. Small lengths of wire can be used as hoops to hold the runners in position, and the pots should contain an all-purpose potting compost or equal parts of loam, peat and sand. Within five weeks or so, the pots will contain a good amount of roots, and the plantlets can then be severed from the parent plant and planted into position for growing on. New stock should not be raised by this method if it is suspected that the old stock is in any way of poor quality – for example, if it does not produce strong flower trusses. It would be far better to buy new plants.

Strawberries require very careful handling by their stalks, and whilst some exhibitors find it possible to secure the fruits complete with stalks by nipping them off with a sharp thumbnail, I think a safer method is to use scissors. The fruits should not look over-ripe when harvested and they should be placed gently into a tissue paper- or cotton wool-lined container and kept in a cool place overnight – it not being desirable to pick them earlier than the eve of the show. As with other soft fruits, they can be laid out onto tissue paper on the staging table for final selection of those that best match in colour, shape and size; and then they can be placed in rows on a plate, stalks to the rear.

A Collection of Fruits

I have listed a few kinds of fruits that should enable a number of entries to be made in a summer show and additionally in a September show. But in a class for 'a collection of fruits' for which there is sometimes a stipulation about number(s) of kinds or cultivars, because of the relative difficulty of producing show-worthy specimens of various kinds of fruit, a table of points values has been evolved, and it is printed in the RHS Handbook. Normally, this system of 'pointing' is used to determine the order of merit in classes for collections of fruit or when there is very close competition making it difficult otherwise to decide between certain exhibits. Similarly there is a pointing system for vegetables and there are tables of points for flowers.

In fruit, there is a considerable range of points values, and in the selection I have suggested there is one 'top-pointer' – dessert apples – and another high-pointed kind – straw-berries. The others I have mentioned are more modestly valued in terms of points. So, if later on there is the urge to enter a collective class in which the competition is fierce, it would be wise to extend the list I have provided by including one or two kinds of fruits that are honoured with a high points value. A melon, for example, is rated at top level (20 points maximum) and is a fruit that can be grown successfully without exceptional difficulties. But, for a beginning, I think that my list is adequate, and it contains kinds of fruit that are normally acceptable for family consumption.

12. A Selection of Vegetables

As with fruit, I have decided upon a short list which I think is adequate for a start, and the kinds of vegetables mentioned are popular in most kitchens. As stated earlier, the physical effort in harvesting vegetables is more demanding than in cutting flowers or gathering some kinds of fruit, but the undermentioned list, with one exception perhaps, does not contain anything of particular difficulty in that respect, nor in regard to cultivation. No doubt, however, the fascination of the challenge of growing massive pot leeks or giant sticks of celery will compel many readers to take up that challenge later on, and so the high standards of National Vegetable championships will be maintained.

Beans
Runner beans seem to symbolise summer in the vegetable plot, and the urge to grow them is apparent in all sorts of situations, some of them far from being ideal. The bean poles of yesteryear are not so evident as are bamboo canes which, despite earlier doubts, I think are a suitable means of support for the vines. Fairly expensive to buy, the canes can fortunately be kept in service for many years if they are lifted from the soil at the end of each season and cleaned with a cloth soaked in Jeyes fluid, dried and stored; and 2.5 m (8 ft) canes pushed well into the ground will take a lot of battering from winds if they are braced to supporting posts or wigwam-style with a horizontally placed cane tied firmly towards the tops of the upright ones. Whether one has a single or double row or some other arrangement will, however, depend upon the space that is available, and clearly the massive iron structures that are to be seen on allotment sites are not appropriate to small gardens. Whatever the method of providing a climbing frame for the vines, it must be firm because otherwise there is a risk that one morning after heavy winds, you might be confronted with the sad sight of the vines lying in muddy soil and looking as if they have no will to flourish any more.

All manner of advice is available concerning spacing the plants in their rows, but bearing in mind that runners enjoy a warm, humid atmosphere, I think that large gaps between plants allowing the wind to dry the soil underneath them are to be avoided, and so about 25 cm (10 in) between them is appropriate. It is also possible to sow between two rows spaced about 60 cm (2 ft) apart, a further row of seeds at, say, rather more than 30 cm (1 ft) intervals, and the plants so produced will add to the denseness of foliage that will help to create a humid environment, provided that plenty of soakings with water are administered. The plants in the centre can, if it is felt desirable, have their growing tips pinched out at about 1.25 cm (4 ft). Whether seeds are sown direct into their permanent stations, or into pots for subsequent transfer to open ground, is largely a matter of personal preference. I find that both methods are equally successful, although one year some marauding pheasants dug up the young plants as they emerged from the soil and a reserve of pot-grown plants had to be substituted. Provided that when planting out the weather is kind, there might be something to be gained by raising plants in pots, as they might produce pods slightly earlier than otherwise and thereby be in time for a particular summer show. Sowing in April in a cold greenhouse or frame would then be appropriate in areas where wintry weather had receded: in the open ground, however, seeds could not normally be sown before May, and in my case, about the third week in May, which usually means that I have some good pods to show in early September.

To guarantee strong, quick growth, beans should be planted in a prepared site which, in effect, means that a trench of at least 30 cms (1 ft) deep should be filled to within a few inches of the top with composted organic material, the top few inches being refilled with the topsoil removed when digging the trench. Fortunately, once this has been done, the site can be used for several years, with top dressings of compost being applied annually. If the effort of taking out a trench is beyond one's resources, the alternative will be to work into the soil compost and peat or a dressing of a proprietary animal manure product, or a calcified seaweed product. During the growing season, liquid feeds of Maxicrop or Phostrogen at about fortnightly intervals will help, and a high nitrogen feed of nitrate of potash, which also provides a welcome supply of potassium, at a rate of about 30 g per metre (1 oz per yd) run will be beneficial if applied in June/July after the plants have established themselves for a few weeks. An old-time favourite alternative is to water the rows with a solution of dehydrated lime in a can of water, using about ½ oz (15 g) of lime in a 1½ gallon (7 litre) can. In a dry season, plenty of water will be required if nice long pods are sought.

Pods should be collected on the basis of straightness, even length of some 25 cm (10 in) at least, and about 2 cm ($\frac{3}{4}$ in) wide, with no obvious sign of seeds; and scissors will be required in order to secure complete stalks. Pods on the inside of the rows will inevitably be pale in colour and will have no appeal on the show bench. The day before the show is quite early enough to collect the pods, and they should be wrapped in a cloth and placed in a cool place. For earlier picking, a moist cloth will help to preserve freshness, or the pods can be wrapped in paper, enclosed in a polythene bag and placed in the crispator of the refrigerator, but they will have to be given a chance of ridding themselves of the condensation that will form. A neat arrangement of pods laid across a plate, tails away from the front of the bench is all that is required (Figure 12.1).

Figure 12.1
Dish of nine beans.
(a) The exhibit is spoiled by one pod without stalk and by an 'old' pod showing the seeds; (b) a correctly-presented dish

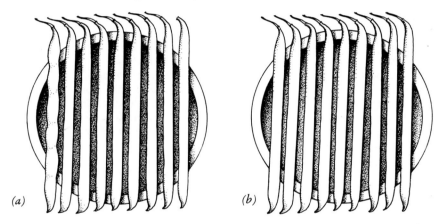

(a)　　　　　(b)

Dwarf French beans offer an alternative of a slightly less highly pointed vegetable for those who cannot cope with, or might not like, 'runners', and they can be grown in largish pots of about 25 cms (10 in) diameter, crocked at the bottom and filled with potting compost. Twiggy sticks will prevent growth sprawling over the sides of the pot, and four seeds per pot sown in late April or May should produce well, particularly if the plants are given regular liquid feeds and not allowed to become dry at the roots. In the open ground, they will produce prolifically if nourished in the same way as for 'runners'. Lengths of twine tied to posts will prevent their flopping onto the ground. Picking and staging is as for runner beans.

As a matter of interest, I should add that the RHS Handbook distinguishes between 'Dwarf French' and 'Stringless' beans as two different kinds of vegetables for show purposes. But it seems that some of the round-podded cultivars are listed in seed catalogues under either or both headings, and it is advisable to resolve any doubts on this score with the show secretary.

Beetroot

Growing long beet to perfection is not at all easy, but the globe cultivars are not over-demanding in their requirements for site and soil conditions, and good crops are easily obtainable from a sowing into a good tilth produced from thorough raking. Roots can be available for summer shows from an April sowing, but germination is poor if the ground is cold; from a May sowing progress is likely to be quicker, although exhibits would not then be ready for an early summer show. The use of cloches to warm the soil first before sowing will, of course, help considerably towards getting specimens for a summer show, and without this aid, one could settle for a less early start and aim for a September show. Obviously, where possible, two or three small sowings from April until the end of May should be made so that it can be determined what is practicable by way of a timetable.

A dressing of Growmore or fish, blood and bone fertilizer along the rows, applying a small handful to about a 2 m (6 ft) run – that is, sprinkling it lightly along, after the plants have been thinned to about 12.5 cm (5 in) apart, should ensure steady growth; and as the shoulders force themselves above the ground, they should be lightly covered by drawing up the soil with a hoe on both sides of each row.

Roots of the size of a tennis ball are quite large enough, and something smaller is acceptable, and lifting them two or three days before the show should do no harm if they are kept in a cool place. The skins can so easily be damaged, so the roots should not be dragged from the soil but helped out with the aid of a small spade placed judiciously to one side and underneath them. Careful washing with a sponge and cold water, preferably under a running tap, is required, and gently rolling the roots in a cloth will dry them. Leaf stalks are usually trimmed to some 7.5 cm (3 in) in length and tied with raffia, but check with the show schedule on this point. The roots can be kept in a dry cloth to prevent damage on the way to the show, and they should be staged on plates, tops away from the front of the bench (Plate 14).

Celery

The growing of sticks of celery of some 1-1.25 m (3-4 ft) in total length is no mean feat, and exhibitors who do this deserve our admiration. At the moment, the RHS Handbook does not differentiate between cultivars of celery in regard to pointing values, and a self-blanching cultivar would qualify for the same maximum total of 20 points as would the usually much larger blanched cultivars. The former could not seriously be expected to vie with well-grown specimens of the latter, and indeed in major championships, it would not be possible to exhibit them in the same class; but often in a local show it will be possible to exhibit self-blanching or American green celery in a class for 'any other vegetable' or in a collective exhibit. I would therefore recommend that the American green cultivar is grown from seed sown under glass with gentle heat in mid-March, the plants being pricked-off into boxes of all-purpose compost as soon as two small true leaves are formed, and the plants subsequently being hardened-off in a cold frame or in a sheltered spot on the patio, with protection from frost at night. At the end of May or in early June, the plants should be set out 25 cm (10 in) or more apart, to such a depth that the roots are covered, but not sunk into trenches. Holes should be dug with a trowel and filled with potting compost. Copious watering will be necessary and slugs guarded against by the usual methods of applying a preventative/killer. Maxicrop at fortnightly intervals – or Phostrogen if that is preferable – or indeed any other suitable liquid feed, including Sangral – should be applied to ensure quick growth of juicy stems, the aim being to produce a thickset head with solid, fresh-looking leaf-stalks.

The leaf stalks will be brittle, and care must be taken when lifting the sticks to avoid breaking them; at the time of harvesting, therefore, the soil needs to be well watered, and a spade should be thrust well down below the crown and to one side and with a firm grasp of the foliage it should be possible gently to ease the stick out of the ground. Roots must be neatly trimmed with a sharp knife so that a slightly cone-shaped stump remains, and all dirt must be removed by washing under a tap or with a gentle flow from a hose, whilst holding the plant upside down. The sticks should be wrapped in a cloth and kept in a cool place, hopefully for just one night, but if earlier lifting is necessary, the foliage should be sprayed with cold water and a damp cloth laid on the stalks. Care must be taken in transporting the sticks to the show, protecting them with a cloth or tissue paper padding, and it is normal in local shows for two sticks (heads) to be exhibited by laying them side by side on the bench, foliage pointing away from the front.

Onions

To produce the mammoth bulbs that one sees at major shows, and also quite often at local shows, requires considerable skill and adherence to a fairly strict time-table involving sowing in late December or in January when, of course, heat will be required. But some good, firm bulbs with a nice golden skin can be produced from 'sets', some cultivars of which will produce semi-globe shaped bulbs which look more attractive on the show bench than the flattish ones. Planted in March or April, the sets or small bulbs will develop into mature specimens by July or August, and it is possible to purchase specially-prepared sets which because of previous heat treatment will be raring to grow as soon as they are planted. It is also possible to purchase sets for autumn/early winter planting which will mature in June or thereabouts in the same manner as plants grown from seed of the 'Japanese-type' onions sown in mid to late August (depending upon location).

Although the bulbs produced from sets are not going to be in the size range of those produced from early-sown seed of mammoth cultivars, similar cultivation measures should be taken if at all possible because the bulbs need to be well-grown in order to obtain their best appearance, and if sets are planted in poor soil, it cannot be expected that they will do well. The 'onion bed' should therefore be well prepared so that it contains a good measure of composted material, including well-decomposed animal manure – but not fresh manure. A good tilth from raking the topsoil should be engineered and a week or so before planting, a dressing of Growmore or fish, blood and bone fertilizer should be raked-in. These days, the sets are planted so that they are just about covered with soil, with the very tips showing above the surface and about 12.5 cm (5 in) apart in rows that are as close as 30 cm (1 ft) to each other. But to allow for reasonable movement between rows without breaking the foliage, I would advise extending the space between the rows to about 40 cm (16 in). In early June, plants grown from spring-sown sets should be given a nitrogen boost by the application of a thin sprinkling between the rows of sulphate of ammonia or nitrate of soda, and a month or so later, a similar dressing of sulphate of potash can be given. If there is no rain about when these fertilizers are applied, a good watering will be necessary.

For a summer show it is customary to show onions 'As grown', that is, with foliage complete, and whereas for those to be shown ripened, the bulbs would be lifted a week or so before the show, it would be inappropriate to do so if the tops are to be retained, so lifting a day or so before the show will be adequate. The bulbs then need to be washed at their bases to remove all dirt, and gently dried with a cloth, and their apparent robustness must not deceive the fact that they can be bruised by rough handling. The bulbs can be laid onto the bench side by side with their foliage stretched out away from the front. If, of course, there is only the one class for onions in a summer show, it could be that you might be competing against an exhibitor who has raised bulbs from 'autumn-sown' seed, and it is then to be expected that they would outweigh those grown from spring-planted sets, but size is not everything, and overall quality will be important. Equally, bulbs raised from a sowing of 'Japanese-type' cultivars or autumn-planted sets would be ripened and the foliage would be getting wizened, and they would look nicer if 'dressed' – that is, with tops shortened and neatly tied; but that could not be done if the schedule asked for 'Onions as grown'. The importance of reading the schedule cannot be over-emphasised. In addition, however, it is quite in order to show bulbs grown from sets in other classes for onion exhibits provided that the schedule does not say otherwise; and a particular class to bear in mind is for 'Onions up to (or under) 8 ozs (220 gms)'.

In order to get onions with a ripened skin, certain measures are helpful. In July, rough skin can be carefully removed from the growing bulbs, and some exhibitors go further than this and gently peel off a layer, but this really does require experience. Two-three weeks before the show, the roots need to have their firm contact with the soil released by thrusting a fork under them and gently lifting slightly upwards, but not enough to lift the bulbs clear of the ground. The idea is to persuade the longer roots to lose their grip, but not to unseat the bulb completely. A week or so before the show, bulbs should be lifted by careful use of a fork and a firm but gentle grasp of the foliage, cleaned by washing under a tap, and placed in a dry, airy place. The tops can be dressed at leisure by reducing them to about 9 cm ($3\frac{1}{2}$ in), bending them down and binding them with some raffia. It is well worth making a neat job of this.

Once at the show, the onions can be staged on rings. Nowadays plastic rings can be purchased, but an inexpensive idea is to cut sections about 2.5 cm (1 in) in depth from the cardboard roll used in kitchen paper and other products, and if the depth is varied somewhat it will enable a group of, say, five onions to be staged so that three sit on the back slightly above the front two, the whole forming a triangle. In staging them in this prominent manner, emphasis is given to 'Uniformity' or lack of it, and as that factor counts for a quarter of the points that may be awarded, it is very important when selecting specimens for exhibiting to choose those that match in size, shape and colour.

A Collection of Vegetables

From the small group of vegetables mentioned, it is possible, in fact, to enter a class for a 'collection of three kinds' in a September show with reasonable prospects, because celery and onions both qualify for a maximum mark of 20 and runner beans for 18, which gives a maximum potential for the three kinds of only slightly less than could be gained if each kind were to be rated at 20 (the highest possible). Whilst at national level, exhibitors would invariably show only top-rated vegetables in a collection where kinds are not specified, it is at least possible to gain, say, 16 points out of a maximum of 18 for a dish of runner beans and outshine some fairly good potatoes gaining '14 or 15 out of 20'. Stretching things a little further, very good beet might win 13 points (out of 15) and score as high as some fairish carrots which are rated at 20 points maximum. The essence of it all is to grow top-quality produce and aim as high as possible. That way, there is bound to be at least some success.

13. Pot Plants

In the majority of cases, a flower show will have classes for 'Pot Plants', and sometimes entries are encouraged by societies' providing at nominal cost items to be grown and shown at a forthcoming show. Bulbs of particular cultivars of daffodils and tulips for growing in a pot of specific dimensions are a favourite item in this connection, and if this is also supplied by the society concerned, there is relatively little excuse for members to escape contributing to the show. Spring shows, in fact, so often have some very nice entries of pots of growing daffodils, and in the case of one or two popular cultivars such as 'Mount Hood' – as already stated in an earlier chapter – 'the same pot' is produced for several consecutive years. Strangely, in my view, some societies preclude this by stating that an award will not be given to a pot of bulbs which was entered and won a prize in the previous year. So be it, but I am not particularly in favour of rules that inhibit rather than encourage entries.

African Violets

African violets (Saintpaulia) are prominent at shows at most times of the year, some people having them in bloom almost continuously! The aim is to produce a plant that has a good symmetrical head of bloom, and a cause of failure is overpotting, so the space between outside extremities of roots and the inside of the pot should not be in excess of 1.25 cm ($\frac{1}{2}$ in), and less rather than more would be preferable. This applies to most flowering plants, particularly fuchsias. Over-watering seems to be a common cause of bud drop, and it is recommended that the pot be allowed to dry out between waterings. The plant should not be placed in a dark position, but it will not relish the sun's rays beating upon it. A humid rather than a dry atmosphere is desirable, and hence it is no bad thing to stand the pot on some small pebbles in a saucer and keep those moist. Or, one pot can be placed inside a slightly larger one, the gap being filled with moist peat (Figure 13.1), but if that is done, the frequency of watering should be held in check. Propagating new plants is

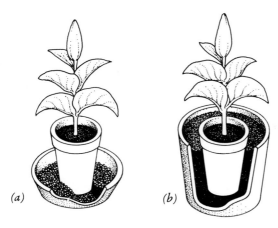

Figure 13.1
Pot plants. Preserv-
ing moisture by:
(a) placing on a tray
of moist pebbles;
(b) packing into a
pot of moist peat

(a)

(b)

fairly easy by inserting leaves by 1.25 cm ($\frac{1}{2}$ in) or so of leaf stalk, so that the base of the leaf sits at surface level, into a pot of equal proportions of sharp sand and peat, or into potting compost, which should be thoroughly moist, and enclosing the pot in a polythene bag. Placed on a warm window-sill, the potted leaf should soon develop roots, and within some weeks it should be possible to pot up the young plant that emerges above the surface, although if a sand and peat mixture has been used, the plantlet will not receive nourishment therefrom and matters must not be left too long. Experimentation will quickly make you an expert in the subject!

Aspidistras

Aspidistras have been regaining popularity, and a well-grown specimen with plenty of shiny dark green leaves should do well in a class for a foliage plant. In fact, they have become valuable, so they should not be left in an unlocked car! They seem to like being somewhat restricted in their pots and to prefer their rhizomatous roots breaking the surface, rather like bearded irises. Despite their rugged and tolerant nature, however, a mild tonic by way of a few drops of Baby Bio or similar product added to the can of water every other time that watering is carried out will help. The leaves collect dust, and this must be removed from time to time by gently wiping them with a wad of cotton wool moistened in tepid water. If a proprietary cleaner is to be used, it might be advisable to give it a polish accordingly prior to the show, although I know that some judges are not impressed by gleaming specimens. Equally, dusty ones will be frowned upon. I suggest that you should get the feel of local custom in this regard, but in any event exhibit a perfectly clean plant.

Begonias

Begonias grown from tubers purchased in the spring, placed in a box of moist peat in a warmish spot under cover, and subsequently when leaves are showing, potted singly into compost in a 15 cm (6 in) pot, make attractive plants for mid-summer, but they are not an easy item to convey to a show because the nodding motion that they may have to endure tends to shake off the blooms. But they are rather exotic-looking and well worth a try.

Busy Lizzie

Busy Lizzies (Impatiens) appear practically everywhere, or so it seems. Giant specimens are grown on office window-sills, but to win an award they must have plenty of flowers, as indeed must all plants entered in a class for a 'Flowering Plant' and, of course, whilst the foliage must not pre-dominate, it must be healthy. They can be grown quite easily from seed sown in the spring, or from cuttings if they can be begged from a friend; or, young plants can be purchased. Aim at stocky, well-jointed plants carrying lots of flowers. Some cultivars will have this character, and it would be wise to purchase seed or plants with such a specification.

Cactus and Succulents

The cactus and the succulent are evident at shows at all times of the year and in a way they represent good 'bankers'; but they should not be regarded as just some item to be trotted-out each time in order to provide a mere entry to help fill the bench. Often the exhibitor is spared the problem of deciding between a cactus and another type of succulent because the schedule will say 'Cactus/Succulent' thus leaving the choice 'wide open'.

Should a cactus be asked for, without qualification, however, the plant should be seen to have areoles, which resemble miniature pin-cushions from which spines or barbed hairs emanate – although in some cases, this will require close inspection. Succulents such as crassula, hoya or sempervivum, would not therefore be admissible. As a matter of further interest, the definition of a succulent is a plant with very fleshy stems and/or leaves, and this includes most cactaceae. Broadly speaking, a cactus showing evidence of flowers is to be preferred to one of similar classification that has no such sign, provided that the former is at least as well grown: but this is, I think, an indication that it would be well worthwhile reading a specialist book on the subject. A cactus on the show bench should invite admiration, and an entry of a small collection, particularly of plants carrying their exquisite and exotic blooms, is, or should be, beautifully attractive.

Fuchsias

Fuchsias gain my personal admiration as beautiful plants of great variety that can be grown and trained in so many ways; and with care and attention that is not over-demanding, they will bloom each year for a long period. Of course, it is nice to contemplate growing a majestic standard plant meeting the specification that the stem should have no side-growths (laterals) until a height is reached of at least 75 cm (30 in) and not more than 105 cm (42 in) from soil level. The head should desirably be equal in width to two-thirds the height of the stem and its depth should be about one-third of that height. A formidable task, and transporting such a giant needs some skill: but it is a challenge worth taking up at some time. To begin with, however, I would advocate growing some good 'bushes', which are plants growing on a short stem of no more than 4 cm (1½ in) and they should have an abundance of foliage and flowers, the latter predominant. Choosing cultivars that are notably floriferous commends itself in this respect of ensuring that an entry in the 'Flowering Plant' class really does have plenty of flowers. Two good cultivars for the purpose are 'Lady Isobel Barnett' and 'Snowcap'. So long as they can be protected from frost, fuchsias will over-winter satisfactorily, and, of course, there are some hardy ones that will spring into new growth each year after dying down in the winter in the open ground. For show purposes, however, we are speaking of pot-grown plants that will be sheltered during the winter under the staging of a greenhouse. The pots should lie on their sides for a while and, if there is no heat available, with a covering of straw or dry peat to afford protection against the pot becoming frozen solid. As always with winter protection of plants that are to lie dormant, little, if any, watering should be carried out, and certainly the compost should not be wet. It really is a matter of trial to determine the best method of winter storage, but in general, damp and cold conditions are a hazard and likely to cause far more harm than a dry, cold atmosphere. A lot more on the subject concerning fuchsias and about training them as top-class plants can be read in a specialist book, and I think it would be worthwhile purchasing one or borrowing one from the local Library; but if the reader is already practising the art of 'pinching out' to produce bushy plants rather than tall, leggy ones, the same idea can be adopted for fuchsia plants purchased from one of the several specialist nurseries in the spring. It is quite possible to build up one's own store of knowledge, including the time intervals between 'stopping' or pinching out the tips of shoots and flowering, but for an assured start, consultation with the 'manual' would be helpful.

Mother-in-law's Tongue

'Mother-in-law's tongue' (Sansevieria) which by its very name can cause a few smirks, can provide a handsome entry in the 'Foliage' class. Its tough, almost leathery leaves should not invite neglect, and careful attention to its growing needs including providing fresh compost from time to time, should be rewarded by the production of fleshy new growths, making a plant of considerable stature all round. It is surprising what a difference there can be between a well-grown specimen and one that only too often is seen on a window-sill in a draughty corridor pleading for love and affection. So, from spring until summer, a few drops of liquid fertilizer should be added to the 'can of water', and the leaves should be gently cleaned of dust at regular intervals.

'Pot et Fleur'

A particularly interesting class in some shows is for a 'Pot et Fleur', which is not quite what a literal translation would require. In flower show terms, it entails exhibiting 'A decoration of growing plants, in or out of pots, arranged in a large container, with cut flowers'. Therefore one needs a glazed container of bowl-like stature, filled with compost (all-purpose type will do nicely) into which plants can be set either in or out of their individual pots, together with a vase of water for holding some cut flowers. The pots should be covered by the compost and the vase should be concealed up to the top of its rim. The advantage of planting with pots intact is that a change of plant is easily effected with the minimum of disturbance to the remainder of the arrangement. The cut flowers should be of modest quantity and in scale with the growing plants.

All pots must be scrupulously cleaned and plants tidied up by removing all dead or dying leaves and flowers; and as this can be done at home, little or no further work should be necessary at the show beyond wiping away any dirty marks incurred in transit and a final check that all spent material has been disposed of. It is important that leaves and flowers are removed carefully, without leaving stubs or stalks showing themselves to the judges. The hardest part, perhaps, lies in transporting pot plants to the show: I cannot pretend that there is an easy answer to this, but hopefully the pots can be packed into boxes and conveyed by car.

Most local societies are generous in accepting plants that have been in members' possession for a relatively short time, and the RHS Handbook states that all exhibits must be the property of the exhibitor and must have been in his or her possession for at least two months, unless some shorter period is stipulated in the show schedule. The schedule therefore requires to be checked on this aspect. On the whole, however, I feel that exhibitors prefer to show plants that they have nurtured for a longer period – perhaps from a cutting or seed; and there is a particular sense of pride if this is so.

As a final word on the subject of pot plants, may I stress that the size of pot must conform to the dimensions stated in the schedule. If it is stated that a pot not exceeding 15 cm (6 in) is required, a mere 1.25 cm ($\frac{1}{2}$ in) in excess will disqualify the exhibit. Furthermore, the schedule may ask for a plant, and that means one plant only; if, as so often happens, the schedule asks for 'A Pot containing a plant or plants' – and sometimes this is done by asking for '. . . Plant(s)' – the exhibitor is free to choose whether to show one or more plants in a pot. Simple perhaps, but so often exhibitors stumble over these points: and how unnecessary and disconsoling that is.

14. Judging

I think it is commonly accepted that judges at flower shows are human beings who do their very best to limit their fallibility to the minimum possible; and usually their experience is such that seldom, if ever, should there be any sense of outrage at their decisions. To be fair to the exhibitors and to arrive at judgements with no personal bias are two requirements met by all the judges of my acqaintance, and whilst a panel of judges may initially not be unanimous in their views on the merit of exhibits in a particular class, particularly if a considerable number of exhibits is involved, invariably discussion leads to a harmoniously agreed decision.

I have referred in earlier chapters to the system of pointing exhibits that exists, and points values will be found, for example, in the RHS Handbook. As shown in chapter 8, it is not only horticultural produce that is sometimes subject to judging on a pointing basis, but there are tables or systems also applicable to the Domestic and Wines sections. But usually the pointing of exhibits is reserved for collective exhibits, such as a collection of several dishes of fruit or three vases of roses, where otherwise it would not be a straightforward matter of deciding the order of merit. It is open to the judges to decide whether to point exhibits, but the time in which the total task of judging a show has to be completed would seldom permit pointing every exhibit; nor would it be necessary.

It is customary for judges to have a preliminary look round the show once all exhibitors have been cleared from the hall or marquee and thus to form an appreciation of the general standard of exhibits. On then reaching the particular section to be judged, the judge(s) will start with the first class in the section and look carefully at each exhibit. This will involve handling exhibits for close scrutiny, and vases of flowers are picked up so that an inspection can be made of the underside of blooms and foliage, some vegetables are cut with a knife or snapped to test for their youthfulness, wine is sampled, and so are bread and cakes and the like, the 'finish' of knitted garments is studied – and so forth. Where there are large numbers of entries in a class, a process of eliminating the notably inferior exhibits will be necessary. A total entry of twelve dishes of potatoes, for example, would need to be whittled down to half that number in a first assessment, and then further examination would determine the three or four of the six and the order of merit of those. Thoroughness combined with reasonable speed is essential: dithering leads to all sorts of problems, not least that the queue of anxious contestants waiting to enter the hall to ascertain how well they have done will have to stay outside beyond the scheduled time for 'opening'.

Judges must respect the fact that the show bench has been arranged to present an attractive spectacle for the waiting public, and it should never be necessary to disarrange everything to such an extent that the stewards have frantically to carry out a major tidying-up operation. Equally, of course, it is to be hoped that the layout will not call for extraordinary measures to be taken by the judges in order to get a close look at exhibits. Of course, the exhibitor places his or her exhibits at risk in entering them in the show, but I have not come upon a case of damage having been done to anything during the course of judging other than the ritual slicing of a beetroot and the snapping of a bean; and the cutting of cakes – which is usually done in such a kind way that they can be auctioned 'as new' after the show, should that be the plan. This point is important to the exhibitor who wishes to take his six onions to another show – a frequent and legitimate practice. I have, in fact, heard of the case where a famous personality in the catering world was asked to judge a show and was not restrained from cutting all produce with a chef's knife. So for once, certain exhibitors must have been highly disturbed to see their choice fruit and vegetables prematurely ready for the stewing pan and certainly not in any fit state for further exhibiting. But, that is a rare case indeed and not at all a likely eventuality.

Very often, it is possible to choose the winner of a class because its merit is obviously outstanding compared with all the other exhibits. It then might be difficult to award second and third places because the virtues of the other exhibits are very low key and their faults considerable. But this is a normal part of a judge's job and it is not right, in my view, to avoid the responsibility of making those judgements by not awarding second and/or third places. Nor, at local shows, should there be an over-abundance of the judge's denoting what they consider to be a wide gap between the best exhibits in a class by perhaps giving a 'first' to one of them and a 'third' to another, with no second prize being awarded. I think that a judge is asked to a show to decide upon order of merit and this should be done; I accept, however, that a standard for a show has to be set and that really sub-standard exhibits should not be awarded prizes as a matter of routine. And, of course, at national shows and the like, one would expect a harder line to be taken or a higher standard set.

A difficult task is to determine the 'Best in Show' should that be unqualified by any limiting factor such as 'Horticultural classes only'. The only possible approach is for the judges to get together and, having decided upon the 'Best' in each section for which they have individually or in small groups been responsible, discuss if any of those has quite outstanding qualities: but it is no mean feat to be sure that the vase of chrysanthemums should beat the strawberry jam or vice versa! Fortunately, more often there will be a 'Best' in each or in some of the sections and perhaps a further award of a Best exhibit in the Horticultural classes, so that there is not the difficulty of assessing the relative merits of entirely different types of exhibits. There are other intriguing possibilities, however, such as choosing the 'Best Rose'. To determine the winner, all the roses would have to be considered, and not solely those that form part of a winning exhibit: a time-consuming task, of course.

These are the lengths to which judges must go in their job, and I think that it is generally appreciated that they do it in a thorough and totally impartial manner. From the exhibitor's point of view, I would suggest, however, that it is worth reflecting upon the impact that is made initially as the judges survey the exhibits in any class, and 'Uniformity' is something that inescapably hits them in the eye, so to speak. In cases where exhibits are pointed, it is a factor that accounts for something of the average order of 25 per cent of points in so far as fruit and vegetables are concerned, and for flowers, many varieties of which are subject where necessary to a pointing system, the comparative percentage is about 15. But, from a visual point of view, the evenness of shape, colour and size or lack of it is something that makes a big impact whether fruit, flowers or vegetables or such items as bread rolls or fancy cakes; and so it deserves special consideration by the exhibitor when selecting specimens to make up a multi-item exhibit. On the same theme, it is very important to present a good first impression of neatness and freshness to the judges, bearing in mind that an exhibit with an unwashed or unclean appearance really deserves to be disliked. Having said that, it must be expected that judges will be far from superficial and will thoroughly examine the exhibits even if their general presentation leaves something to be desired; but one might as well get off to a good start by inviting initial approval for an exhibit that clearly shows itself to be thoughtfully staged.

15. The Show Schedule

The RHS Handbook provides guidance for schedule-makers, and it makes suggestions for avoiding ambiguity. The difficulty is that most schedules of local societies have been evolved over the years by a series of amendments to earlier versions, and seldom can time be found for a complete re-write, with the result that despite correction in certain areas, anomalies remain in others. At times, too, there is a slight reluctance to adopt new terminology introduced by national societies, and I think that this results from a fear that members long used to the old terms will be thrown into confusion by the change. The answer there perhaps would be to make the changes recommended and explain them in notes in the schedule; and this I have seen done quite successfully. Most local societies, in fact, are affiliated to a number of national societies as well as being affiliated in many cases to the RHS, and I think that it would be of general benefit if the local schedules were to reproduce the terms employed by those societies regarding classification, nomenclature and size parameters: in other words, the aim would be 'standardisation'. But I realise the difficulties to be surmounted in this regard.

Some terms must always be free from confusion, a prime example being the use of the words 'Cultivars' (generally acceptable for show purposes to include 'Varieties' and therefore, in fact, representing both terms) and 'Kinds'. In illustration of the two terms, different 'kinds' of flowers are, for example, roses and sweet peas, and different 'cultivars' of them are 'Peace' and 'Noel Sutton', respectively. And similarly, apples and gooseberries are 'kinds' of fruit, with 'cultivars' of those being, for example, 'Blenheim Orange' and 'Leveller'. And with vegetables, for example, onions and potatoes ('kinds') and 'Ailsa Craig' and 'Croft' ('cultivars').

'Must' and 'Should' are simple terms in themselves that appear frequently in show schedules, and whilst I am very much in favour of laying down strictly the rules for exhibitors, there are occasions when the discreet use of 'Should' can avoid the embarrassment of marking exhibits as

131

'Not As Schedule'. Indeed, combined with the word 'About', some red faces can be prevented. For example, saying that 'Leaves must be trimmed to a length of three inches' really means that anything less precise in measurement stands to be disqualified, and this is not likely to be what any society would wish: so, 'Should be trimmed to about three inches' (and/or the equivalent metric measurement) would be preferable. But sometimes 'must' is apt, for example when there are space limitations. It is, therefore, simply a matter of deciding what is intended when writing the schedule and using the term that is appropriate to that intention.

I have already mentioned, in Chapter 10, the difficulty that can arise concerning the use of the word 'Annual(s)', bearing in mind that, unless qualified, it will exclude some of the bedding flowers that we grow as half-hardy annuals but are perennial by nature. The antirrhinum is an example, and often it will be specifically mentioned in a schedule as being permitted to be shown as an annual. But if it is not so mentioned and the schedule asks for a 'Vase of Annuals', it is not an admissible entry. The RHS recommend as an alternative to 'Annuals', the employment of the sentence 'Flowers raised from seed during the twelve months preceding the show'. An alternative, if the term 'Annuals' is to be insisted upon, is to write a note in the schedule saying that the term is to be taken as including all flowers which are raised from seed and normally discarded after flowering after one summer. In either event, sweet williams grown usually as biennials would be admissible, so if it is wished to exclude them and have them shown in a separate class, the schedule will have to say so. Of course, the matter does not rest at the clarification of 'what is an annual', because the use of the word 'Perennial(s)' must also be thought about. As it stands, it includes shrubs and trees, so if the show organisers are seeking exhibits of other perennial plants only, it might be as well to ask for 'Hardy, Herbaceous Plants'.

Accuracy, brevity and clarity are required, but nobody would suggest that they are always achieved simultaneously with ease. I have looked at some show schedules and have taken some examples at random where sometimes ambiguity has been eliminated, but in other cases possibly entries could be made that might not be quite what the show committee had in mind when framing the schedule.

'African marigolds, 6 blooms.'
I think that the society would expect six stems each bearing a large bloom, but a multi-headed stem would not be disqualifiable unless the total number of blooms in the vase was thereby increased to more than six.

'One vase of annuals – three kinds in a vase.'
In this case, the society did state that antirrhinums could be
included, but strictly any other perennial grown as an annual
would be excluded. Only three kinds of flowers are admissible – not more nor less – but any number of stems is
permitted and also any number of cultivars.

'Cacti and/or succulents, 3 pots or pans, different kind in
each pot.'
This is a class with wide scope, permitting a total of three
pots (or pans) each containing one or more plants which can
be succulents or cacti or one of the pots could contain a
cactus and the other two contain plants which are succulents
but not cacti: and so forth! The size of pot is not restricted,
and the only physical limitation is that each pot must contain
a kind of plant, whether it be cactus or other succulent, that
is different from the kinds in the other two pots. I think that
is enough comment except to add that in the RHS Handbook
there is a reference to the use of 'genus', 'species' and
'hybrid' in specialist shows for certain plants, including cacti
and succulents, the terms 'kind' and 'cultivar' not being
sufficiently precise.

'3 large-flowered roses, one or more cultivars, in one
vase.'
I think that the society concerned anticipates three stems
being shown, and this could be made clear by writing:
'One vase, 3 stems, large-flowered roses, one or more
cultivars.'

'3 carrots, long, 3 in (75 mm) maximum tops.'
No ambiguity, but the exhibitor must not let his 'tops'
exceed 3 in (75 mm).

'2 marrows.'
Specific and non-contentious, I feel.

'Tomatoes, 6, red.'
Any cultivar that bears red fruit is admissible, and the plants
can be indoor or outdoor-grown: there is not much room for
argument.

'Jam, soft fruit, 1 glass jar (approx. 1 lb).'
I think that this is nice and clear about what is required and I
feel that the qualifying 'approx.' is particularly helpful these
days when sizes of containers do not correspond precisely to
the old-fashioned ones.

'A soft toy.'
Any sort of soft toy of any size and colour can be exhibited in this class.

'An item in wood.'
Another far-ranging class in the Handicraft section, permitting anything made of wood. Presumably, in judging this, craftsmanship would be of prime importance so that the small beautifully-made model could compete on equal terms with, say, a coffee table.

I hope that in these few examples and assisted by the comments made, I have provided some food for thought without implying that there must inevitably be some way in which the wording of classes in show schedules can be criticised. Most of what is in them is clear enough to exhibitors and judges; and that is what matters. But if time can be found in between shows to go over the wording in your mind and reflect on whether it categorically states what it is that desirably should be exhibited, I think that some useful amendments can often be made to bring about a more satisfying schedule in which the society can rightly take pride as a model of perfection.

16. Summary and Conclusions

I think that it is important to re-state the enjoyment to be gained from participating in flower shows. If the effort completely overwhelms the pleasure, then I think that the balance must be redressed.

Usually, the winning of an award card as a beginner well compensates for the work put in; and hopefully, tired and exhausted committees get ample satisfaction from seeing that yet another show has been a success. Winning at Flower Shows entails not only growing or making an exhibit, but also, in the case of horticultural exhibits, careful harvesting and preparation, and in all cases, safe conveyance to the show venue and neat and attractive staging. A checklist of what is required in general is given here.

Cultivation

An extra touch of loving care is required. This includes a feed of fertilizer for the permanent residents of the garden, such as the apple tree which is often ignored. And basically, a high humus content should be steadily achieved in the soil by working into it plenty of composted organic material. Freedom from pests and disease is important, but indiscriminate spraying must be avoided by a programme that achieves a good clean-up in the winter with perhaps Jeyes fluid coming to one's aid in certain quarters, so that a build-up of trouble is avoided.

Harvesting

Flowers must be given a good drink after cutting and certain steps taken to ensure a good take-up of water. Stems which are hollow should be cut under water to overcome airlocks, and delphinium stems can be filled with water and plugged with cotton wool. Hard and woody stems need to be slit at the bottom or split by gentle hammering, and those producing latex should be cauterised.

Fruit must be carefully handled and all natural 'bloom' preserved. Stalks must be retained. Containers with soft linings should be on site when picking is carried out.

All vestige of dirt must be removed from vegetables. Stalks of beans and peas, as well as on such vegetables as cucumbers, must be preserved intact. Where storage is necessary, cool conditions are required; and damp cloths can be used to keep some vegetables turgid.

Conveying Exhibits to the Show

Careful packing beforehand will save exhibits from damage. Flowers will normally be safe in a flower box, provided that the blooms are well protected by tissue paper or cotton wool and not allowed in contact with wet stems. Fruit and vegetables can rest in their soft-lined containers, and pots can be wedged into boxes; but steps must be taken to prevent their sudden movement in transit. Handicraft items need special care, particularly if glass is involved: this can usually be adequately protected by thick newspaper. Protection from rain will sometimes be an unwelcome but necessary requirement.

Staging

A calm state of mind is most helpful when it comes to staging, and not of least importance is to watch where exhibits are placed in the staging area so that they are neither a nuisance to other exhibitors nor themselves damaged. It follows that late arrival is of no help whatsoever. Probably the best plan is to place the ready-made exhibits in position on the show bench first of all, and then proceed to tackle those that are more time-consuming before moving finally to the remainder. So the cakes and jam can be disposed of, then the tall vase of mixed flowers can be arranged after that, and the exhibits of just one or two stems and so forth can be staged. This, however, is a matter that must remain at the discretion of the exhibitor, and I simply feel that if one has to speed-up at all towards the end of staging time, it is preferable not then to have to do so when dealing with the more 'difficult' objects.

Most important – check with the schedule whilst staging and before committing exhibits to the show bench. This is the final opportunity to be sure that numbers are correct and that all the schedule requirements have been complied with.

Judging

After judging is completed and re-entry to the hall or marquee is permitted, some disappointment may be experienced at seeing that an exhibit has not been awarded what you, the exhibitor, feel should have been the premier position. Careful study of competing exhibits may well prove that there is indeed quite a lot to be said for one or two of them; and the points so observed should be noted for following up when exhibiting next time. As with all pursuits, it is always possible to improve your performance, and determination to do that is more likely to help you overtake the competitor who annoyingly is just that little bit better than yourself than feeling that he or she has just been lucky – although, of course, that might sometimes be the case!

In Chapters 10-13, I have selected items for horticultural classes. The range, I think, is reasonably broad, and some of the flowers are more difficult to grow to a high standard than others; so it is up to the beginner to choose carefully. I did not want, however, to make the list too restrictive; and indeed I suspect that many people are already growing some that require a high degree of skill, but who have so far not ventured into the show world. I hope that they will feel the urge to do so. And, where necessary, I hope that there will be sufficient keenness to obtain specialist books on the cultivation of produce that has special appeal: in a book of this nature I could do no more than given general guidelines.

APPENDIX I
Terms in common use and referred to in this book

Bowl A vessel with a mouth wider than its height, or at least equal to it.

Class A sub-division of a show schedule, there being a class for each group of comparable exhibits.

Collection An exhibit comprising an assembly of kinds and/or cultivars of flowers, fruit or vegetables.

Compost Used to denote composted organic material, such as farm-yard manure, leaves, waste vegetable matter from the kitchen, etc. In sowing seeds, potting-up seedlings and taking cuttings, it represents a specially-prepared mix of loam, peat and sand (or Perlite/Vermiculite, etc.) or a peat-based mix, all with a measured fertilizer element added.

Container This is an almost limitless description of a vessel used to hold water and flowers or fruit and vegetables or various other items. Unlike a bowl or vase, it has no fixed criteria regarding shape and size other than any specific dimensions that might be stated in the show schedule.

Crocked The bottom (of a pot) covered by a layer of broken pieces of flowerpot.

Cultivar For show purposes, this is now gaining acceptance as encompassing cultivated varieties and naturally-found ones.

Dish A specified number or quantity of fruits or vegetables, constituting one item, for showing either as a separate exhibit or as part of a composite exhibit.

Dressing Tidying-up exhibits, particularly flowers, including brushing them gently so that petals assume positions of technical correctness.

Fruit spur A short, stocky growth carrying flower buds.

Humus	The product of composted organic material – usually dark in colour.
Potting-on	Transferring a plant with its rootball intact, or partly so, into a slightly larger pot with the necessary addition of some fresh compost.
Pricking-off	Transplanting small seedlings from the containers in which they have been grown from seed into larger boxes of compost. Once developed, the plants can be put into the open ground.
Set	To set means that flowers have been successfully pollinated to form embryo fruits.
Show bench	Usually a table covered with green hessian or similar material or, in the case of the Domestic section of a show, by a white cloth, and on which the exhibits are staged.
Show schedule	A composite document, usually printed as a booklet, containing all relevant details concerning a show, including rules on eligibility of entry, specification of classes and times of opening and staging.
Soft fruit	Fruit of a soft texture with numerous seeds, e.g. blackberries and strawberries.
Staging	This refers to the preparation of exhibits in an area at the show venue which is usually adjacent to the show hall itself. The term is also used to define the placing of exhibits on the show bench and, in some cases, to describe the tiered structure that is sometimes employed for staging exhibits.
Stone fruit	This has a soft flesh surrounding the comparatively large 'stone' – which is the seed container. Examples are cherries and peaches.
True leaves	The first leaves to appear which are characteristic of the plant.
Vase	A vessel of greater height than the width of its mouth.

APPENDIX II
Useful reference books

Annuals or *Year Books* produced by national societies, and
issued free to members

Catterall, Eric, *Growing Begonias* (Croom Helm)

Edwards, Colin, *Delphiniums* (Dent)

Garrity, John Bentley, *Gladioli for Everyone* (Redwood
Burn)

Gibson, Michael, *Growing Roses* (Croom Helm)

Hessayon, Dr, *The House Plant Expert* and other *Expert*
titles (Pan Britannica Industries)

Jennings, K. and V. Miller, *Growing Fuchsias* (Croom Helm)

Jones, Bernard R. *The Complete Guide to Sweet Peas* (John
Gifford)

Rochford, Thomas, *The Collingridge Book of Cacti and
Other Succulents* (Collingridge)

Royal Horticultural Society, *Concise Encyclopaedia of
Gardening Techniques* (RHS Publications)

—— *The Fruit Garden Displayed* (RHS Publications)

—— *The Horticultural Show Handbook* (RHS Publications)

——*The Vegetable Garden Displayed* (RHS Publications)

Taylor, Jean, *Creative Flower Arrangement* (Stanley Paul)

APPENDIX III
Extract of a typical local show schedule

FIFTY-THIRD ANNUAL SHOW
WELLFIELD PARK, E.13

Prize Money for the 53rd Annual Show kindly
donated by Jolly Garden-Leisure Centre.
Saturday 18th and Sunday 19th August, 1984,
2 pm to 6 pm
Trophies presented at 4.30 pm on
Saturday, 18th August.

Entries for floral art classes to be returned by Tuesday
14th August. All other entries to be returned by Wednesday
15th August. Entrance Fee 5p each Class unless otherwise stated.
Entry forms on pages 40 and 41. Late entries cannot be accepted.

PRIZE MONEY
(Divisions 1 and 2)
Classes 24 and 33: 80p, 60p, 40p.
Classes 11, 46, 54, 57: 60p, 40p, 30p.
All other Classes: 50p, 30p, 20p unless otherwise stated.

Exhibits may be staged on Friday evening 17th August, from
4 pm to 8 pm and on Saturday 18th August, from
8 am to 10.30 am.
NO EXHIBIT MAY BE STAGED AFTER 10.30 am on
Saturday 18th August.

DIVISION 1.—VEGETABLES AND FRUIT
(One variety per Class unless otherwise stated.)

Class
1.—Beans, Runner, 12 pods.
2.—Beans, Dwarf, 12 pods.
3.—Beetroot, 3.
4.—Cabbage, Green, 2 to be shown with stalk.
5.—Carrots, 6 to be shown with 3 in. (76 mm.) of tops.
6.—Celery, 2 heads.
7.—Cucumbers, 2.

8.—**Leeks,** 3.
9.—**Lettuce,** 2.
10.—**Marrows,** a pair for table use.
11.—**Onions,** 6 dressed (not grown from sets).
12.—**Onions,** 6 grown from sets. Dressed.
13.—**Parsnips,** 2.
14.—**Peas,** 12 pods.
15.—**Potatoes,** 5 white.
16.—**Potatoes,** 5 coloured.
17.—**Rhubarb,** 3 sticks trimmed, and tied.
18.—**Shallots,** 12 culinary, large.
19.—**Shallots,** 12 pickling, small, not exceeding 1 in. (25 mm.) in diameter.
20.—**Tomatoes,** 6 red.
21.—**Tomatoes,** 1 truss, outdoor grown.
22.—**Turnips,** 3.
23.—**Vegetables,** any other kind of vegetable not previously mentioned in Schedule.
24.—**Vegetables,** collection of 4 different kinds (see rules).
25.—**Apples,** 3 culinary.
26.—**Apples,** 3 dessert.
27.—**Blackberries,** 15 with stalks, as grown.
28.—**Pears,** 3.
29.—**Plums,** 9.
30.—**Fruit,** any other kind of fruit not previously mentioned in Schedule.

DIVISION 2.—FLOWERS

POT PLANTS
Top of pot or pan not to exceed 180 mm in any direction (inside measurement).

31.—**Cactus or Succulent.**
32.—**Cacti and/or Succulents,** 3 pots or pans, different kind in each pot.
33.—**Flowering Plant,** excluding that grown from a bulb, corm or tuber.
34.—**Flowering Plant,** grown from a bulb, corm or tuber.
35.—**Flowering and/or Foliage Plants,** 3 pots different kind in each pot.
36.—**Flowering, Foliage Plant or Cactus,** specimen plant in pot exceeding 180 mm.
37.—**Foliage Plant.**
38.—**Fuchsia.**
39.—**Pelargonium,** variegated or ornamental foliage.
40.—**Regal Pelargonium.**
41.—**Zonal Pelargonium.**

GENERAL FLOWERS
All items to be staged in a vase unless otherwise stated.

42.—**African Marigolds,** 6 stems.
43.—**Annuals Mixed,** (see rules).
44.—**Asters Single,** 6 stems.

45.—**Asters Double,** 6 stems.
46.—**Fuchsia,** 6 florets, any variety or varieties, (own container).
47.—**Geranium or Pelargonium,** 3 heads.
48.—**Gladiolus,** 1 specimen spike.
49.—**Gladioli,** 3 specimen spikes.
50.—**Herbaceous Perennials,** mixed (Dahlias and early flowering Chrysanthemums excluded)
51.—**Hydrangea,** 3 heads.
52.—**Mixed Garden Flowers.**
53.—**Garden Flowers,** 1 kind not mentioned in this Division.
54.—**Pinks,** 6 stems.
55.—**Rose,** large flowered (Hybrid Tea type), 1 specimen bloom.
56.—**Roses,** large flowered (Hybrid Tea type), 3 specimen blooms.
57.—**Roses,** large flowered (Hybrid Tea type), 6 specimen blooms.
58.—**Roses,** cluster flowered (Floribunda type), 3 stems.
59.—**Roses,** large flowered (Hybrid Tea type), specimen blooms arranged in a Bowl not to exceed 8 ins. (203 mm) in diameter.
60.—**Zinnias,** 6 stems.
61.—**Buttonhole,** Flower or flowers to be dressed as a buttonhole.

DIVISION 3.—FLORAL ARRANGEMENT

Nurserymen, Market Gardeners and anyone engaged in floral work excluded.

Entrance Fee: 5p each Class,
Prize Money in each class: 50p, 30p, 20p.

"SUMMER DREAMS"

62.—**"Cottage Garden".** Novices. Natural plant material. 2ft. 3ins. (686mm). Members only.
63.—**"Joy Of Roses".** Intermediates. An exhibit. Fresh plant material. 2ft. 3ins. (686mm). Members only.
64.—**"All Our Yesterdays".** An exhibit. Natural plant material. 3ft. (914mm). Members only.
65.—**"Garden Party".** A table decoration. Accessories permitted. Fresh plant material. Tablecloth to be provided by exhibitor. To be staged on a table 2ft. 3ins. square and visualised as part of a larger setting. Members only.
66.—**"Gardeners' World".** An exhibit. Natural plant material. 3ft. (914mm). Open.
67.—**"Harmony".** An exhibit incorporating fabric. Natural plant material. 2ft. 3ins. (686mm). Open.
68.—**"Summer Madness".** An exhibit. Abstract or modern. Natural plant material. 2ft. (610mm). Open.

Note The dimensions stated are maxima for height and width.

DIVISION 4.—DOMESTIC SCIENCE AND HANDICRAFT

Prize Money in each Class: 50p, 30p, 20p.

69.—**Bottled Fruit,** including rhubarb, 1 glass jar.
70.—**Jam,** soft fruit, 1 glass jar (approximately 1lb.)
71.—**Jam,** stone fruit, 1 glass jar.

72.—**Marmalade,** 1 glass jar.
73.—**Pickles or Chutney, or Pickled Onions or Red Cabbage,** 1 glass jar.
74.—**Fresh Fruit Jelly,** 1 glass jar (approximately 1lb.)
75.—**Wine,** red sweet.
76.—**Wine,** red dry.
77.—**Wine,** white sweet.
78.—**Wine,** white dry.
    Special note: In the four preceding classes, wine should be exhibited in 26oz. bottles.
79.—**Plate Pie,** any fruit, plate not to exceed 10ins.
80.—**Madeira Cake,** not exceeding 7ins. diameter.
81.—**Chocolate cake,** (own recipe).
82.—**Victoria Jam Sandwich.**
83.—**Decorated Gateau.**
84.—**Savoury Flan.**
85.—**Loaf,** made with yeast. Own recipe.
86.—**Fruit Scones,** 6.
87.—**Shortbread,** 6 pieces.
88.—**Sausage Rolls,** 6.
89.—**Cheese Straws,** plate of 12.
90.—**Knitted Garment.**
91.—**Crocheted Article.**
92.—**An Article of Clothing.**
93.—**Soft Toy.**
94.—**Article of Handicraft.** Pottery, Woodwork, Metalwork.
95.—**Article of Handicraft.** Pictures, (oil or water colour), Collage, Nail and Thread, Embroidered, Dried Flowers.
96.—**Article of Handicraft.** Not mentioned above.

DIVISION 5.—CHILDREN'S CLASSES
Age to be stated on each Exhibitor's card.

No Entrance Fee. Prize Money: 50p, 30p, 20p.

97.—**Pot-et-Fleur.** Growing plants (in or out of pots) and cut flowers (in water or moisture retaining material) assembled in one basic container not to exceed 10ins. (254mm) in diameter. Moss, driftwood, rock and other accessories may be included. Additional cut foliage is not permitted.
98.—**A stuffed toy.**
99.—**Jam tarts,** 6.
100.—**A buttonhole** for a lady.
101.—**A figure or model** made from vegetables and/or fruit.
102.—**Miniature garden** in standard plastic seed tray 10½ins. × 8½ins. (365mm × 215mm).

APPENDIX IV
Notes regarding frost tolerance and sunshine levels

Frost Tolerance

The selection of 'Perennials' listed in Chapter 10 are all 'Hardy' to slightly varying degrees. They are capable of withstanding lower temperatures than the 'safe' levels mentioned below, but this is not predictable.

Pansies will survive several degrees of frost in winter for a week (=USDA Zone 6), provided that the soil is not waterlogged.

Achillea, gaillardias and michaelmas daisies will normally survive hard frost quite well, enduring a temperature for more than a week of 25°F (USDA Zone 6) and for short spells down to 20°F.

Scabious can be just as hardy, but a week at 22-25°F and perhaps shorter spells down to 20°F (USDA Zone 6) are all that could safely be recommended.

Delphiniums given the right strain of seed, or cultivar, for frost tolerance, should survive periods of frost very well; normally down to about 22°F for a week or so (USDA Zone 6).

Sunshine Levels

Preferences for some sunshine as, for example, specified for delphiniums and zinnias, are intended to indicate that the plants do not wish to be hidden from the sun in summer in the United Kingdom, where rarely, however, does the temperature soar high into the 80s°F and then only for a few days at a time. Zinnias, in fact, will do well at a higher range of temperatures (say, around 90°F) if not deprived of water.

For all the perennial plants mentioned, however, including the delphinium, some dappled shade in such hot spells (that is several days at 80°F or more) would be beneficial, should it not be possible to keep the soil very well watered.

The spring-flowering plants – e.g. daffodils and wallflowers – would not, of course, be conditioned for long spells of temperatures exceeding 65°F.

All the 'Annuals' listed prefer sun to shade, but temperatures exceeding the mid-80s°F for any length of time beyond a few days should be considered excessive unless the plants are growing in a wet environment; and sweet peas, in particular, would appreciate light shading during the hours around mid-day when the sun is at its zenith.

Sweet williams will enjoy sunshine and accept short hot and dry spells (up to 85°F).

Roses, of course, if well supplied with water, will do well during sunny spells with temperatures around 80-85°F and they will grow successfully at higher temperature ranges, although blooms will open rapidly.

The only other shrub mentioned – the camellia – is well known for liking dappled sunshine and not fierce sun. Spring-flowering, the temperature maximum in the UK would be around 65°F, but higher temperatures (perhaps in a conservatory) would be acceptable with a moist atmosphere.

The lilies mentioned appreciate their heads in the sun, and in dry conditions temperatures up to the mid-80s°F would be acceptable for a week or so at a time, although shading of the lower half of the plant should be arranged.

The gladiolus is best planted in a site receiving sun from April to September (in UK), but it must not be allowed to dry out at the roots. Provided that the soil environment, therefore, is moist, temperatures up to 80°F plus are acceptable, but for a sustained spell at or above that level, copious watering would be essential to success.

APPENDIX V
Specialist societies

United Kingdom

British Fuchsia Society,
R. Ewart,
29 Princess Crescent,
Dollar,
Clackmannanshire

British Gladiolus Society,
Mrs M. Rowley,
10 Sandbach Road,
Thurlwood,
Rode Heath,
Stoke-on-Trent   ST7 3RN

Daffodil Society,
D. Barnes,
32 Montgomery Avenue,
Sheffield   SH7 1NZ

Delphinium Society,
C.R. Edwards,
11 Long Grove,
Seer Green,
Bucks.   HP9 2YN

International Camellia
  Society,
(UK rep.),
H.J. Tooby,
Acorns,
Chapel Lane,
Bransford,
Worcester   WR6 5JG

National Association of
  Flower Arrangement
  Societies of Great
  Britain,
21a Denbigh Street,
London   SW1V 2HF

National Begonia Society,
E. Catterall,
3 Gladstone Road,
Dorridge,
Solihull,
W. Midlands

National Sweet Pea Society,
L.H.O. Williams,
Acacia Cottage,
Down Ampney,
Near Cirencester,
Glos.   GL7 5QW

National Vegetable Society,
W.R. Hargreaves,
29 Revidge Road,
Blackburn,
Lancs.   BB2 6JB

National Viola and Pansy
  Society,
E. Hazelton,
16 George Street,
Handsworth,
Birmingham   B21 0EG

The Royal Horticultural
  Society,
Vincent Square,
London   SW1P 2PE
(details of other national
societies can be obtained
from the RHS)

Royal National Rose
  Society,
Lt. Col. K.J. Grapes,
Chiswell Green,
St. Albans,
Herts.   AL2 3NR

Saintpaulia and Houseplant
  Society,
Miss N. Tanburn,
82 Rossmore Court,
Park Road,
London   NW1 6XY

United States of America

American Daffodil Society,
Tyner, NG 27980

The American Horticultural
  Society,
Box 0105,
Mount Vernon, VA 22121
(mailing address)
7931 East Boulevard Drive,
Alexandria, VA 22308
(location address)

American Rose Society,
Box 30,000,
Shreveport, LA 71130

Cactus and Succulent
  Society of America, Inc.,
c/o Virginia F. Martin,
2631 Fairgreen Avenue,
Arcadia, CA 91006

The Garden Club of
  America,
598 Madison Avenue,
New York, NY 10022

Marigold Society of
  America, Inc.,
Box 112,
New Britain, PA 18901

National Fuchsia Society,
South Coast Botanic
  Garden,
26300 Crenshaw Boulevard,
Palos Verdes, CA 90274

North American Gladiolus
  Council,
30 Highland Place,
Peru, IN 46970

North American Lily
  Society,
Box 476,
Waukee, IA 50263

Saintpaulia International,
1800 Grand,
Box 549,
Knoxville, TN 37901

Canada

Canadian Gladiolus Society,
1274-129A Street,
Surrey, BC V4A 3Y4

Canadian Rose Society,
18-12 Castlegrove
  Boulevard,
Don Mills, ON M3A 1K8

INDEX